ENVIRONMENTAL RIGHTS

T0174106

Does the concept of 'environmental rights' have any value?

Environmental Rights offers new perspectives on contemporary debates over rights and environmental issues, criticising the traditional *anthropocentric* formulation and the view that it is meaningful to speak of environmental rights as a sub-set of human rights, infringed when individuals experience an environmental quality falling below a recognised standard.

Drawing on key theories of contemporary philosophers and jurists, as well as case reports from judicial decisions in English, European and US courts, the book examines recent developments within environmental law and policy in the United Kingdom and the European Union. Specific rights of the individual are examined – the right to clean air and water, access to information, the right to participate in environmental decisions – as well as practical obstacles to the exercising of such rights, including problems of scientific evidence, high cost of litigation, and legal recognition of environmental pressure groups.

Beginning with an examination of the notion of rights to the environment and to the identification of such rights, the book moves on to describe the arena (town and country planning) in which many of the subsequent debates over rights have originated. The rights discourse is then developed in the context of specific elements of the environment: the atmosphere, water, ionising radiation, land and wildlife, before the final chapter discusses the legal protection currently enjoyed by certain animal species, and examines the idea of *ecocentric* rights.

Christopher Miller concludes that the environment does not lend itself to a rights discourse, but rather to one which stresses the important duties which individuals must assume if environmental threats to human welfare are to be averted.

Christopher Miller is a Senior Lecturer in Environmental Management at the University of Salford.

ENVIRONMENTAL RIGHTS

RIGHTS

Critical perspectives

Christopher Miller

Routledge
Taylor & Francis Group

LONDON AND NEW YORK

First published 1998
by Routledge
2 Park Square, Milton Park, Abingdon, Oxfordshire OX14 4RN

Simultaneously published in the USA and Canada
by Routledge
711 Third Avenue, New York, NY 10017

First issued in paperback 2015

Routledge is an imprint of the Taylor and Francis Group, an informa business

Typeset in Baskerville by Routledge

British Library Cataloguing in Publication Data
A catalogue record for this book is available from the British Library

Library of Congress Cataloguing in Publication Data
Miller, Christopher, 1948–
Environmental rights : critical perspectives / Christopher Miller.
p.   cm.
Includes bibliographical references and index.
Environmental law–Great Britain. 2. Environmental law.
I. Title.
KD3372.M55   1998   97-51719
344.41'046–dc21   CIP

ISBN13: 978-0-415-75723-2 (pbk)
ISBN13: 978-0-415-17064-2 (hbk)

THIS BOOK IS DEDICATED TO T, N AND J, WHOSE EXPECTATION OF – IF NOT A RIGHT TO – NORMAL FAMILY LIFE WAS OFTEN SUSPENDED DURING ITS WRITING.

CONTENTS

ILLUSTRATIONS

Tables

Figures

PREFACE

Many of the traditional academic disciplines were concerned to some extent with the relationship between humankind and its surroundings. Those venerable institutions which have always taught '*natural* philosophy' rather than 'science' could not have foreseen the latter's recent fall from grace any more than the rediscovery of 'nature' as a cynosure of the humanities. In the last decade of the twentieth century, environmentalism could claim to have replaced Marxism as the ideology which mounts the more effective critique of capitalism. But precisely why the paradigm shift occurred when it did, and why environmentalism assumed its current structure, must perhaps await the studies of future historians of ideas.

But even if environmental studies is not yet (unlike many of its practitioners) middle-aged, it is no longer in its infancy. It cannot indefinitely rely upon the tolerance traditionally extended to error, confusion and sloppy thinking when these are attributable to youthful exuberance. By the same token, if it is to lose its reputation as a 'soft science' and to occupy a respected place within the mainstream curriculum offered by the majority of academic institutions, it must define some boundaries, recognise its limitations and identify those components of both its methods and its objects of inquiry which are, if not unique, significantly different from those of its rivals in the humanities and in the physical and social sciences. My exploration of the notion of 'environmental rights' in this book was motivated by a perceived need to direct a critical scrutiny to both components of this phrase.

The 'rights revolution' of the 1960s was largely an American phenomenon, originating in the particular racial conflicts of that country. But Europe was not immune and it was only a matter of time before the confluence of two political movements led to references in the popular, and later the academic, literature to an 'environmental' brand competing in the moral market with the civil, workers', consumers', women's, children's and animal rights with which we were more familiar. Are there reasons, apart from intellectual curiosity, for examining the latest addition to the rights canon?

First of all: perhaps rights have become a 'good thing', of which we are in danger of having too much. Inflation devalues the currency, and a plethora of rights might lead to a depreciation of those which we value most. (This point can hardly be dismissed as trivial during a decade in which we have witnessed 'ethnic cleansing' and other atrocities believed to have been eradicated from Europe half a century ago. The fact that these occurred (just) outside the territory of the European Union is little cause for comfort. Even less is the recollection that the Union, when it eventually tried, failed dismally to muster sufficient political resolve to halt a civil war in which the most execrable violations of human rights were commonplace.) An indifference to this proliferation will increase the possibility of conflict between classes of rights which seek to protect different sets of interests. But the most cogent reason for restraint, I suggest, lies in the political consequences of a belief, whether or not well founded, in the existence of effective rights.

An assertion of the efficacy of *any* class of rights undoubtedly matters if and when it is used to justify a change of policy. It might be argued that self-regulation in the X field is justified because the X rights are a reliable backstop, offering redress for individuals suffering injustice in those instances when their X interests are unlawfully infringed. Similar arguments might be used to defend deregulation by, for instance, repeal of the statutes governing the powers or the budget of the 'X Protection Authority'; they might also be cited in support of the ending of the state monopoly in the provision of X and its transfer to the private sector. But the more such arguments are advanced, the more disinterested observers will require clear evidence of the successful exercise of X rights in the courts; they will want to be assured that there have been some satisfied plaintiffs. Some – but not too many – for there's the rub: the more frequent and widespread the resort to judicial remedies, the more that society must be in need of deeper political responses to the conditions which provoke that dependence. Paradoxically, the polity, wherein the level of rights' violation generates the most voluminous case law, may not be the one in which, in spite of this tempting wealth of empirical data, academics should begin their study. In other words, if the implementation of environmental (like any other) policy entails more than occasional resort to the courts, a political solution to a deeper problem is indicated.

This book is concerned principally with the inhabitants of England and Wales who, as subjects of Her Majesty, do not enjoy the benefits of rights guaranteed by a written constitution but, as citizens of the European Union, are the beneficiaries of a number of rights which have been declared to be 'inherent' in the Treaty of Union. The first two chapters of this book are concerned with the application of the notion of rights to the environment and to the identification of such rights in English and European law; the third describes the arena (town and country planning) in which many of the subsequent debates over rights have originated. These are followed by five chapters which develop the rights discourse in the context of specific elements of the environment: the atmosphere, water, ionising radiation, land and wildlife. The final chapter begins with a discussion of the legal protection currently enjoyed by certain animal species; it then examines the idea of *ecocentric* rights, namely those which are vested in various features (both animate and inanimate) of the environment itself. An attempt to identify the key differences from the *anthropocentric* forms which dominate the earlier chapters leads to conclusions which are unlikely to please rights enthusiasts.

This is not a comparative study. Despite considerable pains, I am conscious that I have not adequately distinguished Scots (and Northern Ireland) from English law. As we are constantly reminded, European Community law and United Kingdom law (whatever that means) are not separate. When I refer to other member states (notably Germany and the Netherlands) and to the rights tradition which flourishes in the USA, I do so for heuristic purposes. I make no claim to be comprehensive; such a claim would be especially hypocritical coming from someone who argues that the encyclopaedic quality of the term 'environment' frustrates the search for conceptual clarity. The choice of the subject matter in Chapters 4–8 is largely a consequence of my earlier research interests in pollution. I'm sure I am no less aware than my critics of the book's lacunae. The built environment (housing, transport and urban infrastructure) is not explicitly addressed; the law related to noise and odour (undeniably sources of nuisance though they may be) consistently fails to engage my interest; the environment as the source of all physical resources, whether or not renewable, is not pursued in the depth which physical geographers might demand. I am conscious that attempts to use international human rights law to secure environmental objects are under-represented; I would simply plead that other, domestic sources of what might have proven to be 'environmental rights' seemed no less deserving of examination.

But I hope that I will not be accused of something that I have come increasingly to observe with suspicion, namely the assumption that, because we all live in the same world, because we rely upon the same physiological machinery to experience that world, we therefore share a common, unique understanding of the term 'environment'. This is not simply a plea for cross-cultural studies. It is a recognition that, even among the inhabitants of an island as small as Britain, there are those for whom the phrase 'environmental problem' means the possible extinction of the natterjack toad; for some it means a fear that their offspring might suffer congenital defects from dioxins emitted by the nearby waste incinerator; whereas for many it refers to the grim reality of daily life in the inner-city estate or to the frustrations endured by commuters. I do not suggest that such heterogeneity of experience can never be embraced in a single volume; my disdain is directed at those who belittle the task. The publishers of books devoted to matters as diverse as risk analysis, modern architecture and geomorphology would appear to see a marketing advantage in squeezing the e-word into their respective titles. Of course, this book is no exception in that respect; but, given my constant refusal to take the word for granted, I hope I may be excused.

When writing about the rights of the individual, it is sometimes necessary deliberately to avoid plural pronouns. Denied 'they' and 'theirs', it is necessary to find another means of subsuming gender. Various attempts at non-sexist alternatives proved either clumsy or confusing; thus 'he' and 'his' are used whenever it is necessary to denote a hypothetical human individual. I regret any offence that this may cause.

In writing this book I have frequently had occasion to consult *Environmental Law* (3rd edn, 1991, Blackstone) by Stuart Bell and the late Simon Ball. This remains the best source of reference on the environmental law of England and Wales; and it was never part of my intention even to attempt to produce a rival. However, reported cases constitute my largest single source of 'data'. It is here, I believe, that empirical evidence of the effectiveness of the protection afforded by rights is most readily gained. I have tried to create, by analysis of the relevant law and policy, a coherent perspective spanning a number of environmental sectors which are more often treated in isolation. If I have failed, I leave it to the reader to judge whether this is due to my inability to apply sufficient clarity or to the inherent diversity of my chosen sectors. Either way, the fault is mine alone and none can be attached to the many friends and colleagues who, often without knowing it, have assisted me in writing this book. Special thanks are due to George Liebmann, Jim McLoughlin, John Salter, David Hughes, Bill Howarth, John Alder, Ludwig Krämer and Martyn Day for their patience over the years in answering a physicist's interminable questions on law.

ACKNOWLEDGEMENTS

I acknowledge the kind permission of the Controller of Her Majesty's Stationery Office for permission to reproduce extracts from various documents; of British Nuclear Fuels plc (in respect of Figure 6.3) and of the Massachusetts Medical Society (Figure 4.2). I thank the Institute of Advanced Legal Studies and the publishers of the *Journal of Legal Studies* (Blackwell), *Environment and Planning A* (Pion) and the *Journal of Planning and Environmental Management* (Carfax) for permission to quote liberally from articles in which various themes in this book first appeared. I also acknowledge my gratitude to the Nuffield Foundation, whose generous award of a Social Science Research Fellowship in 1995 afforded the respite from teaching duties which made the writing of this book possible.

TABLE OF STATUTES

TABLE OF EUROPEAN
DIRECTIVES

TABLE OF EUROPEAN
REGULATIONS

TABLE OF ARTICLES OF TREATY OF ROME

(as amended)

TABLE OF CASES

England and Wales

European Court of Justice

European Court of Human Rights

USA and CANADA

1

THE CONCEPT OF AN 'ENVIRONMENTAL RIGHT'

Preamble

Living on the eastern seaboard of a vast and sparsely populated continent replete with natural resources, it is unlikely that the draughtsmen of the Constitution of the United States of America gave any thought to environmental constraints which might prejudice the enjoyment of the rights to 'life, liberty and the pursuit of happiness'. It was the familiar Old-World threats – corrupt judges, dogmatic clerics, autocratic kings – to what are now labelled *human* rights which they were most anxious to address. When, nearly two hundred years later, the United Nations declared

> Men and women of full age, without any limitation due to race, nationality or religion, have the right to marry and to found a family.[1]

the cited constraints on the enjoyment of this right were social constructs rather than physical factors. Given the proximity in time of the Holocaust and the experience of many atrocities committed during the Second World War, the absence of any reference to ecological constraints in the *Universal Declaration of Human Rights* is hardly surprising. It was only with the emergence of the environmental movement in the 1960s that the inalienability of the right to reproduce was questioned:

> To couple the concept of the freedom to breed with the belief that everyone born has equal right to the commons is to lock the world into a tragic course of action.[2]

By the time of the 1972 (Stockholm) United Nations Conference on the Human Environment, the idea that an acceptable environment might constitute a precondition for the enjoyment of certain human rights no longer seemed controversial:

> Man has the fundamental right to freedom, equality, and adequate conditions of life, in an environment of a quality that permits a life of dignity and well-being.[3]

Fifteen years later and following the next conference of a similar stature, environmental quality had acquired the status of a 'fundamental' human right:

> All human beings have the fundamental right to an environment adequate for their health and well-being.[4]

The declaration which emerged from the United Nations (Rio de Janeiro) Conference in 1992 was couched in less explicit terms:

> Human beings are at the centre of concerns for sustainable development. They are entitled to a healthy and productive life in harmony with nature.[5]

But the full blooded language of 'rights' is amply apparent in the (1994) Draft Principles on Human Rights and the Environment produced by the UN Sub-Commission on Prevention of Discrimination and Protection of Minorities. The first two of its five parts are reproduced in Box 1.1. Many of these are what Raz[6] has described as 'derivative' (as distinct from 'core') rights; whether the General Assembly will accept so extensive a 'wish list' remains to be seen. If, as the Sub-Commission proposes, these principles are to form a new and distinct addition to international law, a sceptic might point to the mutual antagonism between many of the listed rights and to the enormity of the task facing any state which strives to pay them all more than lip-service.

International law was originally concerned with the eradication of war as a means of resolving disputes between sovereign states. The environment is the most recent concern of international law, although many (if not most) past conflicts were essentially 'environmental' insofar as they involved disputes over land or other resources. It is now possible to point to the charter, conventions and treaties of various international organisations which, although primarily concerned with the protection of human rights, have acquired, or can be interpreted to have, an environmental role. The Council of Europe's Convention for the Protection of Human Rights and Fundamental Freedoms will be discussed further below; the 1988 protocol to the American Convention on Human Rights includes an article which confers a right to 'live in a healthy environment and to have access to basic public services'.[7] Moreover, similar rights are now to be found in the constitutions of nation states in both the New World (Ecuador, Peru, El Salvador and Brazil) and the Old (Spain and Portugal). Within Europe, discussion of environmental rights must be concentrated upon those associated with the treaties setting up the communities (especially the European Economic Community) which now form constituents of the European Union. Chapter 2 explores the irony of a community, originally set up with the undeniably economic aim of creating a single market in goods and services, becoming the source of a bewildering variety – social, political and environmental – of rights.

BOX 1.1 Extract from the United Nations' 'Draft principles on human rights and the environment'

Part I

1 Human rights, an ecologically sound environment, sustainable development and peace are interdependent and indivisible.
2 All persons have the right to a secure, healthy and ecologically sound environment. This right and other human rights, including civil, cultural, economic, political and social rights, are universal, interdependent and indivisible.
3 All persons shall be free from any form of discrimination in regard to actions and decisions that affect the environment.
4 All persons have the right to an environment adequate to meet equitably the needs of present generations and that does not impair the rights of future generations to meet equitably their needs.
5 All persons have the right to freedom from pollution, environmental degradation and activities that adversely affect the environment, threaten life, health, livelihood, well-being or sustainable development within, across or outside national boundaries.

Part II

6 All persons have the right to protection and preservation of the air, soil, sea-ice, flora and fauna, and the essential processes and areas necessary to maintain biological diversity and ecosystems.
7 All persons have the right to the highest attainable standard of health free from environmental harm.
8 All persons have the right to safe and healthy food and water adequate to their well-being.
9 All persons have the right to a safe and healthy working environment.
10 All persons have the right to adequate housing, land tenure and living conditions in a secure, healthy and ecologically sound environment.
11 (a) All persons have the right not to be evicted from their homes or land for the purpose of, or as a consequence of, decisions or actions affecting the environment, except in emergencies or due to a compelling purpose benefitting society as a whole and not attainable by other means.

(b) All persons have a right to participate effectively in decisions and to negotiate concerning their eviction and the right, if evicted, to timely and adequate restitution, compensation and/or appropriate and sufficient accommodation or land.

12 All persons have the right to timely assistance in the event of natural or technological or other human-caused catastrophes.

13 Everyone has the right to benefit equitably from the conservation and sustainable use of nature and natural resources for cultural, ecological, educational, health, livelihood, recreational, spiritual and other purposes. This includes ecologically sound access to nature.

Everyone has the right to preservation of unique sites consistent with the fundamental rights of persons or groups living in the area.

14 Indigenous peoples have the right to control their lands, territories and natural resources and to maintain their traditional way of life. This includes the right to security in the enjoyment of their means of subsistence.

Indigenous peoples have the right to protection against any action or course of conduct that may result in the destruction or degradation of their territories, including land, air, water, sea-ice, wildlife and other resources.

Source: UN Commission on Human Rights, Sub-Commission on Prevention of Discrimination of Minorities, Human Rights and the Environment, Final Report of the Special Rapporteur (UN Doc. E/CN.4/Sub.2/1994/9).

But the increasing role of international organisations prompts an ontological question: at what point is it meaningful to speak of the existence of an environmental (or, for that matter, any other fundamental) right: when it is first declared by a body like the United Nations; when it is translated into national law; or only when that law has been found to offer an effective remedy after that right has been infringed?

Those approaching the question from a common law tradition are predisposed to associating rights with remedies which emerged from 'hard cases'.[8] Most trials are adversarial in character, with one side anxious to refute the relevance, validity and extent of any right claimed by the other. A right which, whenever tested, persistently fails to live up to expectations will, regardless of its pedigree in international law, either lose that designation or fade into oblivion. A close study of contested claims will assist an understanding of the (as yet unrealised) potential of a rights approach within environmental protection. Empirical research alone cannot identify that quality which elevates a demand, preference or interest to the status of a right. If, as some philosophers argue, it is the autonomy[9] of each individual to act according to his volition which is celebrated in and protected by human rights, what is the environmental equivalent?

Given the coincidence, particularly in the USA, of the 'rights revolution' and the growth of environmental consciousness, some level of fusion was perhaps inevitable. It is not difficult to postulate various circumstances in which individual autonomy is restricted, not by the deliberate actions of a repressive state, but by a lack of environmental resources. It may be possible to define a set of conditions C which are necessary for the existence of rights R; but it does not follow that C themselves must be rights. Once R are accorded the status of rights, they are less subject to the vagaries of the normal political process. But that status must be jealously guarded; to invent phrases like 'derivative' or 'second or third generation' rights serves to diminish the importance of the core. If some minimal environmental quality, together with a basic standard of health, peace and social order, is a precondition of the enjoyment of some more fundamental right to human fulfilment, is there a need to speak of *environmental rights*?

'Green' consciousness means that the environment is no longer seen simply as the stage on which the human drama is acted out, but more as an author of the plot. But is it possible to identify a set of environmental rights which are *sui generis* in that they are not derivative of a set of human rights? Can the use of Occam's razor – penetrating the rhetoric and conceptual extravagances which too often characterise popular and, it must be admitted, academic discussion of our chosen subject – leave us with a category of rights which are quintessentially environmental insofar as none of the other familiar epithets is applicable?

Even without its many metaphorical uses, the breadth of the term 'environmental' is daunting. It is now routinely applied to our physical surroundings in the home and at work, to the water we drink and to the food we eat, the air we breathe and to the ozone layer in the stratosphere. It refers also to an awesome variety of plant and animal species. Many would criticise this enumeration of parts as reductionist, for it overlooks the interdependence of the animate and inanimate constituents of this planet. Gaianism[10] represents an extreme version of 'ecologism' which views the Earth as a single, enduring organism, capable of adapting to the varying stresses, whether internal or external, it has faced in its four-billion-year existence.

Although not quite as expansive in its application, the notion of a 'right' is not unproblematic and continues to pose problems for philosophers and lawyers alike. Hohfeld's[11] identification of four distinct concepts subsumed within lay understanding of the term 'right' was of seminal importance: 'claims, privileges, powers and immunities' represent forms of legal relationship, each having its correlative concept: 'duties, no-rights, liabilities and disabilities' respectively. Whilst discussion in this book is centred most often upon claims and duties, it is not difficult to identify examples of the other three correlative pairs with clear 'environmental' connotations. The influence of Hohfeld's legacy is such that it is easily forgotten that the link between correlative pairs is not axiomatic. Because it is often possible to identify the person(s) on whom a duty is conferred by virtue of the right borne by another, it does not mean that it is illogical to imagine an individual bearing a duty with no corresponding right bestowed upon another.

Given the breadth of meaning attached to the concept of a 'right' and to the word 'environmental', the combination might be thought to present a level of indeterminacy more than sufficient to defy academic analysis. In fact, two distinguishable conceptions are beginning to acquire a currency in the literature.

First, it tends to be applied to a broad range of legal rights associated with use of and access to those many elements of the physical world which determine human health and well-being. Of course, an adequate diet is the primary and ultimately environmental prerequisite for that most fundamental human right, namely the right to life. The fact that discussion of human rights has tended to be conducted with little or no regard to the reality of human physiology is perhaps testimony not only to the rarity of famine in industrialised societies, but also to the alienation of the great majority of the population of such societies from the exigencies of food production.

The second conception of environmental rights is ecocentric rather than anthropocentric; these are not human rights in the sense discussed above, but rights – above all the right to a continued existence unthreatened by human activities – attached to non-human species, to elements of the natural world and to inanimate objects. As O'Riordan[12] was the first to point out, the recognition – that an animal, plant or ecosystem should receive the protection which rights can offer – remains an inescapably human construct. No matter who or what is admitted to the 'moral community', it is only the human members who can attach meaning and value to membership of that community.

A synthesis of these two conceptions will be attempted in the final chapter. But first it is necessary to engage in a search for examples of rights, regardless of their particular pedigrees, being successfully invoked in the pursuit of a range of environmental objectives. Even if this investigation fails to isolate a sub-set of definitively environmental rights, our understanding of the more general category will have been improved. Before setting out, it is necessary to declare the prejudices and preconceptions which influence the choice of the parts of the environmental map to receive the closest investigation.

Prejudices and preconceptions

First, I believe that our search can profitably begin by examining what economists call 'common property resources'. The twenty years since Lord Scarman wrote 'English law reduces environmental problems to questions of property'[13] have seen sufficient counter-examples to cast doubt on the current validity of his assertion. However, it is still necessary to ask why it is that property law remains appropriate for some elements of our immediate surroundings (e.g. buildings, riparian rights) whilst the notion of ownership sits so uneasily on others (e.g. the atmosphere, the electromagnetic spectrum, the human genome). Indeed, Professor MacCormick[14] chose to cite 'clean air in a city' when searching for an example of the antithesis of an 'individual good individually enjoyed'.

A concern for the environment has been a major influence in the increase in public law actions. Curiously the recent economic history of the USA and the UK has been one of a shift from the public to the private spheres. Although they will be discussed in later chapters, I am not referring here simply to the recent floatations on the UK stock market of a number of utilities (which had been state-owned for fifty years or so) but to a general reduction of the role of the state in the provision of services in health, welfare, transport and education. Too recent a focus must be avoided. The enclosures of the eighteenth century and earlier are among the first and the most visible manifestations of an historical process of appropriation which left the few remaining instances of 'commons'[15] in England – shared grazing rights in the New Forest; the 'free (coal) miners' of the Forest of Dean; the Purbeck quarrymen in Dorset – as rare curiosities to delight antiquarians.

Sustainability – the shibboleth of contemporary environmentalism – had its origins in economists' responses to tragedies inherent in discrete commons, like fisheries. As the notion of 'Spaceship Earth'[16] became popularised, sustainability came to be concerned with the rate of exhaustion, on a global scale, of all non-renewable resources. It acquired a moral dimension when it questioned the assumed privilege of developed states to consume resources without reference to the potential needs of undeveloped and developing economies (predominantly in the southern hemisphere). It is in its recognition of the needs of generations yet unborn that sustainability raises the most challenging ethical questions.

Sustainable development has been defined as that which 'meets the needs of the present without compromising the ability of future generations to meet their own needs'.[17] Various interpretations have been attempted: ranging from strong sustainability, with onus on the diligent preservation of all species and of non-renewable resources, to the weak variant in which the depletion of the latter can be contemplated provided that compensation, possibly by man-made capital, ensures that overall human welfare does not decline with time. Since the Rio Conference in 1992, the implementation of sustainable policies at national and local level has been a recurrent theme in UK environmental policy. In carrying out its responsibilities, the Environment Agency (and its Scottish counterpart) are statutorily obliged to have regard to the 'objective of attaining sustainable development'.[18] And the latest revision of the document[19] which sets out the broadest aims of the town planning system places sustainable development at the very core (see Chapter 3). Although the practicalities appear daunting, the idea of attributing rights to future generations is not unduly problematic: English courts of equity have for centuries recognised, as legal entities, trusts which exist for the benefit of heirs and successors in perpetuity.

Second, I suspect that any right on which the environmental label sits comfortably will bear some tangible relation to the physical world. By analogy, the biological fact that giving birth is the exclusive prerogative of women, and the social fact that the principal role in nurturing children falls to women in almost all societies, represent constraints (namely reproduction and childcare) on the autonomy of woman. These constraints are sufficient to distinguish a category of 'women's rights'. This sub-set of human rights remains very broad – embracing the right to terminate pregnancy, the abolition of discriminatory practices in employment, to a mother's entitlement (in the UK) to be the sole recipient of child benefit. The category of environmental rights seems likely to be broader still.

To those who argue that conflating the terms 'physical' and 'environmental' is no more than a tautology, two pleas in mitigation might be offered. It serves as a reminder that increasingly common phrases like 'social environment' are metaphors which only confuse our search. More to the point, tautologies have their uses. The evident circularity of two dicta, which will be cited again in this book, has not prevented them from becoming extremely influential in their respective areas of the English law. In regard to land use planning, Lord Scarman has held that 'a planning purpose' is one which is related to 'the character of the use of the land'.[20] Of somewhat wider application and arguably more central to English administrative law, the *Wednesbury* test has been described as tautologous, because it 'allows the courts to interfere with decisions that are unreasonable and then defines an unreasonable decision as one which no reasonable authority would take'.[21]

This declaration of initial preconceptions cannot neglect the author's understanding of the notion of the state and whether it can be said to have an environmental role. According to one American theorist, the state exists in order to fulfil three essential functions:

> First, it defends the basic needs and interests of those who control the means of production within the society in question. Closely associated with this is the second function of the state: achieving legitimacy for itself and ensuring social harmony. This function involves the remarkable fact, brooded upon by political theorists since time immemorial, that somehow the state – which is organised force – becomes an agent of moral authority, and its rule is accepted in the main by those subject to it. Finally, no state can survive if it cannot adequately defend itself, and the dominant powers in the economy and society, from external attack.[22]

Although the above was not written from an explicitly environmental perspective, its view of the state offers a convenient starting point for our discussion. If any 'Tragedy of the Commons' is to be averted then, the state must 'control the means of production' using its 'organised force' to appropriate land, raw materials and sources of energy which, unregulated, could be over-exploited to the point of exhaustion. The notion of 'external attack' can be extended to embrace environmental threats: global warming and a rise in sea levels threaten the very existence of some Pacific island states. However, invasion by foreign powers – as traditionally understood by this phrase – has often been prompted by environmental or resource constraints. The burning of Kuwait's oilfields at the end of the Gulf War was labelled 'environmental warfare' when employed by the retreating Iraqi forces. More importantly, ensuring continued access to oil supplies remains one of the principal determinants of the foreign policy of most developed nations. Denial of so important a resource could be viewed as a hostile act warranting a military response. Whilst resources, especially non-renewables like coal and oil, might become a greater source of conflicts between states, it is necessary to recall that the maintenance of an apparatus for resolving interpersonal disputes, and ensuring that honouring of contracts is the norm, is one of the fundamental roles of the state.[23]

As a product of the 'welfare state' myself, there is one further prejudice which I must declare: it is my predisposition to view the state as essentially benign. I offer this personal statement simply because it is at odds with the credo of the UK politician who dominated the decade of most relevance to this book. At the height of her power, Margaret Thatcher implied that society did not exist; in retirement, she has warned of the state's great potential for evil.[24] Rather than strengthen the rights of individuals to secure redress against abuses of state power, her response to this threat, when in power, was to reduce the role of the state. Nevertheless, the environment represented one area of state responsibility which arguably increased during her period in office.

Rights or remedies?

Since such diverse concerns – from neighbourhood noise to relations between nation states – fall within its remit, it is hardly surprising that environmental law lacks what one English academic lawyer has described as 'a coherent set of principles and rights'.[25] I believe that the lack is not of principles or rights but of coherence. And while it does not indicate an unqualified acceptance of Lord Wilberforce's dictum that 'typically, English law fastens, not upon principles but upon remedies',[26] I defer discussion of principles to the following chapter whilst listing here some of the more important remedies which have been employed, in either private or public actions in English courts, to pursue environmental objectives.

Tort: the original environmental remedy

Many authors of legal textbooks see tort as the *fons et origo* of environmental law. The environmental label sits easily on the majority of cases in trespass and nuisance, simply because they are related to a greater or lesser extent to land use. A very recent action in nuisance, although ultimately failing to remedy a singularly modern source of annoyance, nevertheless offers a very instructive lesson on the limits of this tort.

It was established[27] in 1865 that nuisance can be characterised by two types: damage to land which directly reduces its economic value; and various forms of disturbance which diminish the quiet enjoyment of the land. With the demise of heavy industry in the UK, the former is now the rarer, but damage to crops from pesticides carelessly applied on neighbouring land might be taken as an example. Far more common are manifestations of the second type: impairment of the enjoyment of land as a result of 'emanations' – noise, vibrations, smoke, fumes, dirt, etc. – from another's property. The majority ruling in *Hunter v. Canary Wharf*[28] stressed that these remain two variants of a single tort: in both cases the value of land is reduced, and that reduction will form the starting point for assessing the compensation payable in any successful action for damages. In the second, the loss lies in the 'amenity value', in other words, the monetary valuation which the occupier places upon the satisfactions to be gained from the peaceful enjoyment of his property undisturbed by noise, smoke, fumes, etc.

Having described their understanding of current law, the majority of the House then explained its aversion to arguments seeking to modify it in two distinct ways. Firstly, interference to television reception caused even by an exceptionally tall (250m) building was deemed not to constitute a nuisance. Citing precedents dating from the seventeenth century, it was agreed that the interruption of view and other essentially passive effects of one building upon another were not, in the absence of some easement or covenant, unlawful. Under current UK law, television reception was but one of many 'material considerations' (see Chapter 3) to be taken into account, along with any representation made by local residents[29] to the local planning authority before planning consent was given.

The second ruling in this case confined the right to sue in nuisance to those with a legal title (predominantly freeholders and tenants). The opportunity to extend the right to other occupants (spouses, children, relatives, lodgers, etc.), although clearly recognised by the majority and vigorously advocated in the minority speech, was rejected. Contrary decisions by the Court of Appeal in this case and in another[30] concerned with telephone harassment were overturned.

If industrial noise disturbs the sleep of the ten residents of one house as well as the single occupant of the adjoining dwelling, the aggregated disutility in the first may well be ten times greater than that occurring in the latter. The loss of amenity value is, *ceteris paribus*, the same in both cases since, as the House was anxious to reaffirm, nuisance is a 'tort to the land'[31] and not a tort to the person. It followed that only those with a legal title had standing to sue in nuisance. Even if the health of an owner of land (or a tenant) were to suffer as a result of exposure to noise, dust or some other emanation, any damages awarded in nuisance would still reflect the loss in amenity value of the land and not the health impairment. This point and the restricted standing to sue represent obvious limits to the effectiveness of nuisance as a remedy where health damage is alleged to have arisen from toxic substances dispersed into the atmosphere.

The tort of negligence is based not upon land but on 'fault'. Harm arising from a careless act can be compensated provided the plaintiff can show that the harm was foreseeable and that it arose from a failure of a 'duty of care' owed him by the defendant. Perhaps the major limitation on tortious liability assuming a greater role in modern environmental cases is the extent to which the plaintiff's onus of establishing causation – demonstrating that the balance of probability suggests his harm is caused by the defendant's action (or inaction) – has tended to become a contest between rival scientific experts. Environmental cases typically involve evidence given by epidemiologists and physiologists concerning the long-term consequences of the consumption of chemical and biological contaminants.

There are numerous cases involving harm to health and injury in the workplace which have resulted in successful claims of negligence and breach of statutory duty by employees. In many of these cases, e.g. crush injury by falling loads, amputation by unguarded machinery, *res ipsa loquitur*; others, such as bladder cancers among workers in the rubber industry exposed to beta-naphthylamine,[32] have hinged upon judicial assessment of uncertain epidemiological evidence. Successful claims tend to be those in which in a specific toxic chemical can be readily identified as the causative agent of an uncommon pathological condition; 'specificity of cause' (see Chapter 6) is rarer in environmental cases. It is necessary to pay some attention to one important action in negligence[33] in which compensation would appear to have been (belatedly) awarded for 'environmental' exposure to what is usually seen as an occupational hazard.

It undoubtedly required the dangers of asbestos to be fully and unequivocally recognised before two plaintiffs from Leeds who, although never employed in industries using asbestos, could eventually secure compensation for mesothelioma (cancer of the pleura) – a condition for which exposure to asbestos fibres is the only known cause. If one examines the report[34] of epidemiologists into deaths from mesothelioma reported between 1971 and 1987 in Leeds, then, in the absence of direct or indirect[35] occupational exposure, residence in the vicinity of the J. W. Roberts factory in Armley is of clear importance. Processing of asbestos began at this factory in 1899; a study of the health of Armley workers in the 1920s helped to identify asbestosis as an occupational disease and hence to bring about the first Asbestos Regulations in 1931. Eye-witness reports of the amounts of asbestos (which one of the plaintiffs, as a child, formed into 'snowballs') in the factory yard and in the streets in the immediate vicinity of the factory suggest that the usual difference (as a result of atmospheric dispersion) between occupational and environmental exposures did not arise in this case.

Asbestos litigation worldwide now involves prodigious sums of money and is reported to have contributed to the problems which Lloyds of London encountered in the early 1990s. The case of *Margereson and Hancock v. J. W. Roberts Ltd* is undoubtedly important and paves the way for many other claims by non-occupational victims of asbestos-related diseases. However, some caution should be exercised before this case is held up as an example of a successful use of tort to secure redress for environmental pollution. The circumstances of this case were such that the factory owner's duty of care was adjudged by Holland, J. to extend beyond his employees to nearby residents: 'in the immediate vicinity of the premises, factory conditions in terms of dust levels were at various points effectively replicated so as to give rise to like foresight of potential injury to those exposed for prolonged periods'.[36] In the subsequent appeal, this line of reasoning was accepted and the Court of Appeal added that the risk of 'some pulmonary injury'[37] to children and other residents was foreseeable in 1933 (the date from which Mr Margereson was known to have regularly played with the dust).

That there may be occasions when residents in the close proximity of a factory also receive compensation for a breach of the duty of care, owed primarily to those employed within that factory, is not surprising. Little heuristic purpose is served by imputing an environmental dimension to that body of law concerned with health and safety at work, which can be more usefully understood within the context of the rights which exist between employer and employee. Had the outcome of the *Budden* case[38] (see Box 1.2) been different, then the tort of negligence might have furnished us with a much clearer example of an environmental right – to breathe clean air – being successfully exercised in a UK court.

BOX 1.2 Children's health and lead in petrol: the *Budden* case

Parents of two children living near an urban motorway took action (as next friends) in negligence (a claim in nuisance was struck out at first instance) against BP Oil in respect of the alleged impairment of their children's intellectual development due to the effect of lead additives in petrol. Curiously, the epidemiological evidence for and against this claim appears to have received little scrutiny in open court: Megaw, L. J. is reported to have said that 'it had to be assumed [sic] that the children had suffered some injury caused in part by the fact that the companies' petrol included lead'.[39] The blithe assumption of the validity of the applicants' claims to health damage is quite remarkable given the enormous amount of research which was necessary before the negative effect of lead compounds on the psychological development of children became accepted.[40] However, the case was to depend upon a constitutional point. The Court of Appeal's reason for dismissing the action was that an injunction forbidding the addition of lead compounds to petrol would in effect nullify the statutory emission limit then in force.[41] According to Megaw, L. J.:

> The oil companies could not be held to be negligent and failing in their duty to the children in complying with the requirements prescribed by the Secretary of State and approved by Parliament. If they were liable and to be restricted by injunction to limits lower than those laid down in the regulations the same would apply in all other actions which would doubtless follow so that the courts would be laying down permissible limits inconsistent with those prescribed by Parliament. That would result in a constitutional anomaly which would be wholly unacceptable. The authority of Parliament must prevail.[42]

This decision has been criticised: an immunity from civil actions is nowhere to be found in either the regulations or s.75 of the Control of Pollution Act 1974 which empowered the Secretary of State to make them.[43] The defence of statutory authority is available to highway authorities and the Civil Aviation Authority in regard to noise nuisance from road and aircraft noise respectively. The companies producing and selling petrol were not doing so in fulfilment of some statutory duty.

The 'foreseeability' of the threat to the health of nearby residents posed by the asbestos accumulations was the key to the plaintiffs' success in establishing the negligence of the owners of the Armley factory. Two years before this case, fore-seeability had been declared by the House of Lords[44] to be a condition of the rule in *Rylands v. Fletcher*, which had hitherto been assumed to impose strict liability for certain forms of environmental damage.[45] Nuisance would appear to be further eclipsed by negligence. However, the *Cambridge Water*[46] case (see Chapter 5) has cast doubt upon the future role of tort in environmental disputes generally.

Successful actions in tort in respect of chronic health detriment to third parties from pollutants dispersed into the environment, under conditions which fall short of accidents, remain conspicuous by their paucity. It might be argued that academic reviewers place excessive emphasis on reported cases, and that the true efficacy of tort in such cases lies in the apprehension of large damages which encourages out-of-court settlements. However, such settlements – which may be just and even economically efficient – serve to vitiate the establishment of prece-dents; they are therefore inherently antagonistic to the emergence of legal rights.

Irrespective of the current role of tort, it was largely a response to its inability to combat the consequences of industrialisation and urbanisation that prompted the proliferation of statutes designed to mitigate the conditions which threatened the health of residents of urban centres (who from 1851 formed the majority of the UK population). Of course, 'sanitary reform' necessitated the curtailing of a number of freedoms, such as the choice of the fuel to be burnt in the domestic hearth,[47] which the English had long considered their birthright. Nevertheless, this extension into the statutory arena meant that a wider range of other rights became available in the pursuit of environmental objectives.

Rights and immunities in criminal proceedings

Under English common law, individuals may initiate proceedings in the criminal courts. They need not have suffered as a result of the offence nor need they demonstrate any particular interest. This point is important, for it enables environ-mental organisations to initiate proceedings. This ancient right is subject to the discretionary power of the Attorney General to terminate proceedings. The crim-inal provisions of environmental statutes sometimes contain additional, specific restrictions of the general right of individuals: for instance, the consent of the Attorney General was required before individuals could prosecute offenders under the Rivers (Prevention of Pollution) Act 1951; in private prosecutions, under the Alkali Act 1906 and later the Health and Safety at Work Act 1974, in respect of industrial air pollution, the consent of the Director of Public Prosecutions was required.

A general right (power), which is restricted rather than facilitated by statute, seems an unlikely source of a new species of environmental rights. Why, as authors of textbooks invariably do, restrict discussion to the rights of private individuals to prosecute – what of the rights of those accused of environmental crimes? Again, the claim-rights (to use Hohfeld's terminology) of *habeas corpus* and trial by jury apply to anyone charged with such modern offences as depositing controlled waste on land not subject to a waste management licence,[48] but attaching the environmental label to these ancient common law rights illustrates the fatuity that an uncritical use of the term can entail.

It is possible to point to one time-honoured defence, specific to pollution law, which might be said to constitute an environmental right (an immunity). A demonstration of the use of 'best practicable means' (BPM) to prevent the emission of smoke was a sufficient defence for early Victorian industrialists charged under the Smoke Nuisances (Metropolis) Act 1853. As Ashby[49] points out, BPM began as a defence but by degrees became the foundation of UK pollution control for more than a century until supplanted by BATNEEC (best available techniques not entailing excessive cost; see Chapter 4). BPM still survives as a defence in summary proceedings in statutory nuisance.[50]

But these are minor points when compared with the basic principle of English law that an individual enjoys the privilege of doing anything that is not specifically proscribed. It is hard to disagree with McLoughlin[51] that pollution control laws have been drafted to protect the interests of those causing pollution rather than those suffering its effects. The traditional system in the United Kingdom was not what is today described as 'technology forcing'. Once the technology constituting 'best practicable means' was in place, and a sufficiently tall chimney erected to disperse the residual emissions which defied BPM, it was the operator of a scheduled process who effectively enjoyed a right (namely immunity from prosecution).

The increased rights of access to information (see below) conferred by the Environmental Protection Act 1990, could well encourage more frequent private prosecutions under this Act (which, it should be noted, does not include a provision requiring the consent of the Attorney General or Director of Public Prosecutions). The prospect of the payment of costs has always been a powerful disincentive to private prosecutions. Whether the increased availability of information is viewed by private individuals or, more pertinently, environmental pressure groups as sufficient to overcome the cost deterrent when contemplating actions against multinational chemical companies remains to be seen.[52] Greenpeace successfully took action against Albright and Wilson, in respect of its illicit discharges to the Irish Sea from its Whitehaven plant; the chemical company was fined £2,000 and ordered to pay ten times that amount in costs.[53] It could be that, concerning atmospheric and aquatic discharges, easier public access to data coupled with increased media attention will force the regulatory agencies to pre-empt private intervention by pursuing their own prosecutions and placing less reliance upon extra-legal persuasion.

Statutory modification of tortious liability

Within the context of the notion of the 'right to pollute', it is necessary to make passing reference to the defence of 'statutory authority' in civil actions against a public body with a statutory power to perform an act, such as the disposal of waste, in which nuisance is an inevitable consequence. If reasonable care is taken, then there is no liability.[54] Much depends upon the wording of the empowering statute; if it is permissive, then immunity from the consequences of the action cannot be assumed. Rogers[55] suggests that a 'more rational distinction' depends upon the freedom in the location of the activity: if the statute requires a polluting activity in a particular place, no liability for nuisance arises; but if an undertaker has a choice of location, there is a sense in which he chooses his neighbours and must therefore be conscious of their sensitivities when pursuing any polluting activity. It must be added that where that choice is contingent upon the decision of a local planning authority (see Chapter 3), that authority is not subsequently liable in nuisance.[56]

Occasionally, the balance of public and private interests is recognised as requiring the restriction of the right to pursue certain civil actions. Civil aviation could not have developed if the noise arising from each and every takeoff and landing could occasion action in nuisance or trespass from any resident living near an airport; and the Civil Aviation Act 1982 explicitly removes the right of action in these circumstances.[57] There is provision for affected individuals within certain areas to receive grants towards the cost of sound insulation measures;[58] but anyone aggrieved at the invasion of his privacy by aircraft noise from major airports (and traffic noise from highways[59]) would appear to have no remedy in law (see below).

The realities of modern transport systems may have required the extinction of rights in tort, but it is also necessary to note that the exigencies of nuclear power have led in theory (but see Chapter 6) to their strengthening, to the extent that operators of nuclear installations are strictly liable[60] for damage attributable to radiation from any nuclear material on or released from their plant.

Rights to information

Until very recently the English public has enjoyed no right of access to information held by pollution control agencies. These bodies tended to view any information they might acquire as commercially confidential; rival manufacturers might be able to penetrate various 'trade secrets', given a knowledge of the chemical composition of discharges to the atmosphere or to watercourses. Indeed, unauthorised disclosure could result in prosecution under the Official Secrets Act 1911. Under the Rivers (Prevention of Pollution) Act 1961, s.12, officers of the river authorities were expressly prohibited from disclosing any data obtained from any sampling exercise or from an application for consent to discharge to surface waters. If the discharger consented or if disclosure was required in any criminal proceedings, then the prohibition was lifted.[61] These restrictions persisted during the brief lifetime of Part II of the Control of Pollution Act 1974, in which the regional water authorities were however required to maintain public registers of certain information on discharge consents.

The Royal Commission on Environmental Pollution repeatedly called for a change in official attitudes. Its *Tenth Report* recommended 'a presumption in favour of unrestricted access for the public to information which the pollution control authorities obtain or receive by virtue of their statutory powers, with provision for secrecy only in those circumstances where a genuine case for it can be substantiated'.[62]

The Environment Agency is required to maintain a register of consents to discharge under s.190 of the Water Resources Act 1991. Within the Environmental Protection Act 1990, s.20 confers a right of public access to information on integrated pollution control (IPC), whilst s.78R offers an equivalent in respect of contaminated land. An industrialist may seek to exclude information from the IPC register on the grounds that it is 'commercially confidential'; and the right of appeal exists for those aggrieved at the refusal of this exemption.[63]

The duty of various agencies to maintain registers and the public's right of access to them are not particularly novel; the right to consult registers of planning applications is now fifty years old. The move towards a general duty on public authorities with environmental responsibilities to supply information on request is undoubtedly a landmark, and one for which credit must be shared between the Royal Commission and the European Commission.[64] Commercial confidentiality is listed among the exemptions from the duty of disclosure imposed on public authorities by the Environmental Information Regulations.[65] When these regulations were scrutinised by a House of Lords select committee,[66] the grounds for exempting information from disclosure was one of five areas of concern. A more serious shortcoming was associated with the status of the privatised utilities. If, as the Water Services Association argued,[67] the water companies, as private bodies, fall outside the remit of the regulations (and the original directive), then the public's right of access to information on the most widespread forms of pollution (see Chapter 5) in the United Kingdom is confined to that held by the regulatory bodies – the Drinking Water Inspectorate and the Environment Agency. Friends of the Earth and other environmental pressure groups take a different view, pointing out the extent to which the state continues to exercise control over the 'statutory undertakers', and citing a European Court ruling (see Chapter 2) that a body like British Gas is regarded (at least for the purposes of employment law) as 'an emanation of the state'.[68]

If the UK were ever to follow the United States and pass a 'Freedom of Information Act', many (but by no means all) of these issues might be resolved. In such a regime, the environment would form one, but only one, sector for which access to information had been placed on a statutory basis; thus a guarantee of access to an even greater range of data on emissions, concentrations and impacts would still not amount to a quintessential 'environmental right'. However, this is not to deny that the freer availability of information, especially the results of any monitoring of emissions to atmosphere or watercourses, could assist private individuals intent upon pursuing a private prosecution against polluters or seeking a judicial review of any authorisation by a minister or a pollution control authority.

Judicial review

Cases with a clear environmental dimension are prominent within the spate of judicial reviews which has been so distinctive a feature of recent (post-1990) legal history in the United Kingdom. Greater availability of relevant information cannot alone account for the rise in popularity of this remedy. Of equal if not more importance was the relaxation in the High Court's understanding of the 'sufficient interest'[69] which an applicant for judicial review must be able to demonstrate.

The earlier, restrictive interpretation is best illustrated by the ruling in *Rose Theatre Trust*,[70] in which development on the site of one of the more important theatres in Shakespeare's London was opposed by a group of actors and scholars concerned with Elizabethan drama. This group challenged the Secretary of State's refusal to protect the site by using the powers conferred upon him by the Ancient Monuments and Archaeological Areas Act 1979. When dismissing its application, Schiemann, J. held that this group did not have 'sufficient interest' and, more curiously, he seemed to imply the existence of a category of decisions in which a minister exercises his discretion in the interests of the public as a whole, but which are immune from legal challenge by any group or individual within that public.

Given the preconceptions recited earlier, I am reluctant to include this case – despite the regrettable loss of part of our cultural heritage and the opportunity for archaeological study and possible restoration – within the environmental canon. It is not difficult to point to analogues with a clearer claim: the loss of any habitat of an endangered species (see Chapter 8) is an obvious example. The more localised the threat, the more likely that some individual or amenity group will be adjudged (even by the earlier interpretation) to have 'sufficient interest'. Undeniably the environmental members of Schiemann, J.'s anomalous category are those in which global pollution threatens a global population. We all contribute to, and are affected by, greenhouse warming and ozone depletion, but no single individual can lay claim to a special status (over any other) to challenge any ministerial decision which is perceived to exacerbate such problems.

Subsequent chapters will describe in detail judicial reviews of environmental decisions which have arisen after *Rose Theatre Trust* and which assisted the emergence of a more generous attitude to the standing of non-governmental organisations in the UK. These chapters will refer to a number of instances of clear errors by English judges in cases in which standing was complicated by European law requirements. However, the growing literature[71] points to an emerging consensus on the 'qualifications' which might now be expected to establish the standing of an environmental group:

- objector at a planning (or similar) inquiry
- consultee over white paper, select committee or Royal Commission report, etc.

- relevant scientific expertise (e.g. Royal Society for the Protection of Birds)
- *bona fide* interest (see *Greenpeace (no. 2)*[72] in Chapter 6)

Environmental concerns fell within those 'public interests' which were considered in a report[73] of the Law Commission in 1994. Among its proposals was the setting up[74] of a 'discretionary track' alongside the existing procedure, which would allow the High Court to recognise the standing of individuals acting purely in the public interest to apply for judicial review. This particular recommendation, although not yet enacted, has received the endorsement of the Master of the Rolls in a document[75] which elsewhere declares that 'the growth of public law and, in particular, judicial review has been one of the most significant developments in the English legal system in the last 25 years'.[76] It is worth noting therefore that environment (unlike medical negligence and housing) was not given its own chapter in *Access to Justice*. Moreover, only one explicit reference[77] to the environment is to be found in this report, which is a little surprising since its author can hardly be accused of environmental short-sightedness.[78]

Human or environmental rights?

An individual's right to enjoy land and property in which he has a lawful interest is, as we have seen, fundamental to the tort of nuisance. Privacy is central to domestic felicity; invasion of privacy entails a 'loss of amenity' of the home for which compensation might be payable. The fact that circumstances which an English common lawyer instinctively labels 'private nuisance' can also engage the attention of one concerned with human rights is not particularly surprising. Persistent noise, vibration or odour can prevent an individual from enjoying the basic comforts which his home might offer; it is possible to imagine circumstances in which they amount to no less an invasion of privacy (and are no less conducive of stress) than telephone harassment, the constant presence of 'stalker'[79] or persistent staring by a malicious neighbour. In *Canary Wharf*, the House of Lords was determined that nuisance should not become a backdoor route for the creation of a tort of harassment.[80] More generally, is there an argument for maintaining a clear distinction between activities like telephone-tapping or interference with mail and invasions of privacy by noise, smoke or fumes as a result of the carelessness, incompetence or plain lack of neighbourly consideration? In the former, the state itself is often the agent of the intrusion which it justifies in terms of threats to national security; in the latter, if the state is involved, it is in its failure to institute means for resolving private disputes or enforcing adequate standards. It is not clear that either protection of the environment or the defence of human rights is assisted by glossing over essential differences.

If, as in *Lopez Ostra*[81] (see Box 1.3), judicial reasoning in a human rights case is ultimately based upon an assessment of a 'fair balance' between conflicting land uses, then it is replicating the role of the national civil court; it becomes, in effect, another court of appeal. Of course, it is possible to conceive of a decision so

18

manifestly unbalanced – so skewed against the individual – as to violate any notion of reasonableness; and a human right is undeniably violated if a national jurisdiction fails to overturn it. That conclusion, I suggest, should not be reached lightly by a court which is outside that national jurisdiction. Indeed, this would appear to be the view of the court in a later case (cited again in Chapter 3) which involved a gypsy woman whose caravan home fell foul of local planning policy:

> By reason of their direct and continuous contact with the vital forces of their countries, the national authorities are in principle better placed than an international court to evaluate local needs and conditions. In so far as the exercise of discretion involving a multitude of local factors is inherent in the choice and implementation of planning policies, the national authorities in principle enjoy a wide margin of appreciation.[82]

In the light of this recent ruling in the *Buckley* case, it requires a very careful reading of the report of *Lopez Ostra* to identify the reasons why the European Court of

---

BOX 1.3 *Lopez Ostra v. Spain*

The family home of the applicant in *Lopez Ostra v. Spain* was situated some 12 metres from one of several tanneries in the town of Lorca. The municipality had provided the land and also a subsidy for plant for the treatment of water contaminated in the tanning process. Although the appropriate licence from the local council had not been obtained, an attempt was made to commission the plant, whereupon odoriferous emissions from the plant necessitated the evacuation of residents in the immediate vicinity for three months. On the advice of health officials, the council ordered certain parts of the tannery, but not the water treatment plant, to cease production.

In October 1988, the applicant argued in the Spanish courts that the council's refusal to demand the closure of the water treatment plant had violated two rights under the Spanish Constitution, namely

Article 18: the right to private life and inviolability of the family home.

Article 45: everyone shall have the right to enjoy an environment suitable for personal development and the duty to preserve it.

Eventually the claim – that fumes, odours and noise from the plant entailed a violation of the applicant's rights – was rejected by the Constitutional Court.

In 1990 Mrs Lopez Ostra took her case to the European Commission of Human Rights, arguing that the local council's inaction had infringed her rights under the Convention. A claim that she and her family had been

---

subjected to 'degrading treatment' (Art. 3) was dismissed. The Commission was prepared to refer, to the Court, the argument based upon a violation of Article 8, namely

> 8 (1) Everyone has the right to respect for his private and family life, his home and his correspondence.
> 8 (2) There shall be no interference by a public authority with the exercise of this right except such as is in accordance with the law and is necessary in a democratic society in the interests of national security, public safety or the economic well-being of the country, for the prevention of disorder or crime, for the protection of the rights and freedoms of others.[83]

The Court ruled that there were circumstances in which pollution, although falling short of serious threat to health, could justify action under Article 8:

> Naturally, severe environmental pollution may affect individuals' well-being and prevent them from enjoying their homes in such a way as to affect their private and family life adversely, without, however, seriously endangering their health.[84]

In regard to pollution of this kind, Article 8 places two duties upon public bodies: to take 'reasonable and appropriate measures' to ensure respect for homes and family life; and to prohibit public bodies themselves from causing pollution which interferes with domestic life in this way. When deciding whether a state has breached either duty

> regard must be had to the fair balance that has to be struck between the competing interests of the individual and of the community as a whole, and in any case the state enjoys a certain margin of appreciation.[85]

The various items to be weighed in the balance in this case included the measures taken (temporary rehousing) to protect family life, the fact that the council had itself subsidised the cost of the plant, the fact that the treatment plant was needed to treat effluent from tanneries which were an important part of the local economy. After consideration, the Court ruled that a fair balance had not been reached, the margin of appreciation had been violated and that a breach of Article 8 had arisen. Compensation of about £20,000 was awarded to the applicant.

Human Rights, after considering the balance of individual and community interests, concluded that the Spanish courts had exceeded their 'margin of appreciation'. Two issues appear to be decisive: the evidence of Mrs Lopez Ostra's daughter's paediatrician on the health effects of the emissions; and the fact that it was accepted that the plant was in violation of the local environmental regulations (even though this violation was not deemed by the Spanish courts to amount to a breach of constitutional law).

In the earlier case of *Powell and Rayner v. United Kingdom*,[86] the Commission, when faced with a similar task of finding a fair balance between individual and community interests, came to a different conclusion. Both applicants lived in locations which, to different extents, were subject to noise from aircraft using Heathrow Airport. It was held that aircraft noise, although an indirect and unintended intrusion, could amount to a violation of the right to a person's private life and home (Article 8.1). On this occasion the Commission held, after taking account of the airport's annual £200 million contribution to the balance of payments and the employment it offers to nearly 50,000 people, and the grants for sound insulation totalling £19 million, the UK government's margin of appreciation, implied in Article 8.2, had not been exceeded.

BOX 1.4 *Zander v. Sweden*

The applicants were the owners of land near a waste tip. Cyanide in the leachate from the tip contaminated the groundwater to the point that, when the applicants were no longer able to consume water from the wells on their land, the local authority supplied them with drinking water. However, this arrangement ceased when, in June 1984, Sweden's National Food Agency recommended that the maximum permitted concentration of cyanide in drinking water be raised from 0.01 to 0.1 milligrams per litre. Two years later, when the licence for the tip came to be renewed, the landowners demanded that any renewal should be conditional upon the tip operators supplying them with drinking water free of charge. The Licensing Board dismissed the applicant's claim, and their subsequent appeal to the Swedish Government was similarly dismissed on 17 March 1988. An act which enabled the judicial review of certain administrative decisions, including the appeal against the licence conditions at issue in this case, came into effect on 1 June 1988. Since this Act did not have retroactive effect, it could not be relied upon in this instance. The applicants were effectively denied the right to appeal against a decision of a state body which directly affected their ability to drink water from a well on land which they owned. This was held by ECHR to infringe their rights as landowners, and these property rights were clearly 'civil rights' within the meaning of Article 6.1. The Swedish Government was ordered to pay each plaintiff about £3000 in compensation as well as costs, which were considerably greater.

It is necessary of course to distinguish between the situations in which the national courts arrive at a 'balance' which unreasonably disfavours the individual, and one in which the individual is denied the opportunity to protect his rights. The right of a national remedy for the violation of other rights is contained in Article 13. A second claim in *Rayner and Powell* based upon this article was deemed admissible by the Commission and was referred to the Court.

Section 76 (1) of the Civil Aviation Act 1982 prevented them from taking actions in trespass or nuisance in respect of noise from aircraft flying at a reasonable height and in accordance with Air Navigation Orders made under that Act. They therefore claimed that they were denied an effective remedy before a national authority in respect of the complaints about aircraft noise. This claim was also rejected; it was held that Article 13 was not to be interpreted as requiring signatory states to institute modes of challenge to the legality of their legislation.

Article 6.1 of the Convention – 'In the determination of his civil rights and obligations everyone is entitled to a hearing ... by [a] ... tribunal' – is most often cited in respect of criminal cases. It was successfully invoked in the (civil) case of *Zander v. Sweden*[87] (see Box 1.4). The official report of this case bears the subtitle 'Access to a court for environmental rights'.[88] Although the origins of this dispute were undeniably environmental, Sweden's failure – to institute a means of challenging an appeal against a licence – was essentially a procedural one; it was not contingent upon the content of the licence (water quality) or the particular grounds of the appeal.

Lopez Ostra, despite my reservations, would appear to have a better claim than *Zander* to membership of the 'environmental' club. Both cases were concerned with very localised incidents of pollution; but if attention passes from Europe to the Human Rights Committee of the United Nations, it is possible to identify attempts to use international human rights law to address much wider environmental issues. Protestors argued that French nuclear testing at Muroroa atoll in the South Pacific threatened their right to life and their right to a family life. The Committee dismissed the action,[89] arguing that the applicants were not 'victims' (in what has become the accepted sense for human rights cases) even though it reiterated its view that nuclear weapons testing remained a significant threat to life.

The blurring of the distinction between human rights and environmental protest is a recurring theme in this book. Members of various judiciaries in the United Kingdom may be faced with attempts to use rights arguments to further environmental objectives if and when the Human Rights Act 1998, which incorporates certain rights of the European Convention, takes effect. Insofar as past experience described in the following chapters may be taken as relevant, environmentalists would be unwise to place too much confidence in this strategy.

2

HAS THE EUROPEAN UNION CONFERRED 'ENVIRONMENTAL RIGHTS' ON ITS CITIZENS?

Introduction

In 1977 a joint declaration[1] by the institutions of the European Community demonstrated their desire for the rights provisions of the European Convention on Human Rights to be considered as part of the Community's legal framework. The Treaty already contained what might be termed 'political rights' covering voting, and eligibility to stand, for the European Parliament; and the right of free movement of workers, services and students was entailed in the very notion of a 'common market'. As the Community's interests extended so its citizens acquired 'social rights' covering health and safety at work, equal pay and treatment between men and women, and pregnancy and maternity rights.[2] More recently, six international environmental organisations, including Greenpeace and Friends of the Earth, failed to persuade the Intergovernmental Conference (held in Amsterdam in June 1997) to amend the Treaty of Union so as to include in Article 8d:

> Every citizen of the Union shall have the right to a clean and healthy environment, access to the decision-making process, information, and justice as part of a general right to human development.[3]

This juxtaposition of a 'right to a clean and healthy environment' with rights of access to 'decision-making . . . and justice' is, as we have seen in Chapter 1, not particularly novel. This chapter considers what such a right might mean, and examines the capability of the institutions of the European Union and of its member states to confer it upon their citizens.

The word 'environment' is not to be found in the original Treaty of Rome by which the European Economic Community (EEC) came into being. Article 2 of the 1957 Treaty included among the tasks of this Community 'a continuous and balanced expansion' and 'an accelerated raising of the standard of living'. However, one might argue that objectives, which by today's uncritical usage would be considered environmental, were implicit in the 'other' 1957 Treaty of Rome which set up the European Atomic Energy Authority (Euratom) and, given that coal is a non-renewable resource, in the 1951 Treaty of Paris which created the European Coal and Steel Community.

By 1970 the Council of the EEC had adopted Directives on noise[4] and polluting emissions[5] from motor vehicles. In the absence of Treaty provisions concerned explicitly with the environment, these were promulgated under the authority of Article 100, which removes barriers to the internal market by the approximation of laws. In 1972 the Commission announced its belief in the need for a 'Community action programme on the environment'. In the same year, agreement had been reached on the accession of the United Kingdom, Eire and Denmark to the Community, and therefore it was nine heads of government who declared:

> Economic expansion is not an end in itself. Its first aim should be to enable disparities in living conditions to be reduced. . . . It should result in an improvement in the quality of life as well as in standards of living. As befits the genius of Europe, particular attention will be given to intangible values and to protecting the environment, so that progress may really be put at the service of mankind.[6]

Despite such rhetoric, environmental directives continued to represented in terms of removing barriers to free trade and harmonisation of laws. It was only when the Single European Act in 1987 added Articles 130r,s and t (comprising Title VII 'Environment') to the EEC Treaty that the Community's environmental legislation – now comprising nearly two hundred directives through which the policy objectives of five Community Action Programmes were implemented – was placed upon a legitimate and unequivocal basis. The Treaty of Maastricht 1992 reinforced this basis by ensuring that an unashamedly green concept – 'sustainable and non-inflationary growth respecting the environment' – was cited within the very aims[7] of the European Union (of which the Community is the principal element).

Some disappointment was expressed in some quarters that a form of words[8] closer to that of the Brundtland Commission 1987 was not used. The revision agreed at the Amsterdam Summit (June 1997) – 'balanced and sustainable development of economic activities' – seems unlikely to assuage all the earlier criticism. Nevertheless, the citation of 'sustainable development', on a par with economic objectives, in the very aims of the Union is eloquent testimony to the growth of environmental concerns in the two decades following the Community's first action programme. However, the Single European Act of 1987 represented, for the mechanics of environmental policy, the more important breakthrough; Maastricht built upon these secure foundations. The environmental concerns of Maastricht received far less attention than its wider economic and social objectives. Were it not for the depth of UK government's antipathy to the Community's Social Chapter, culminating in the celebrated 'opt-out'[9] at Maastricht, then perhaps the Westminster Parliament's acceptance of principles as alien to the traditional

British approach as the 'precautionary principle' and the presupposition in favour of 'high levels of protection' might have been less easily obtained. These principles of Community environmental policy were in addition to those such as 'polluter pays', 'prevention at source' and the 'integration' of an environmental dimension in all Community policies which, articulated in the first four Action Programmes, were given substance in the Single European Act. Article 130r(1–4) of the Treaty of Rome 1957 (as amended) is reproduced in Box 2.1 on page 26.

If the celebrated opt-out preserved UK's freedom to determine its social policy without interference from Brussels, the much vaunted principle of 'subsidiarity' was represented as a safeguard against unwarranted intrusion in other policies:

> In areas which do not fall within its exclusive competence, the Community shall take action, in accordance with the principle of subsidiarity, only if and in so far as the objectives of the proposed action cannot be sufficiently achieved by the Member States and can, therefore, by reason of the scale or effects of the proposed action, be better achieved by the Community. Any action by the Community shall not go beyond what is necessary to achieve the objectives of the Treaty.[10]

In fact, subsidiarity had applied to environmental matters since 1987 but there is little publicly available evidence of its being actively advocated by British representatives before its extension to all areas of European Union law in 1992. Air and water pollution have little respect for national boundaries and as one commentator has observed, the environment 'perhaps more than other policy area requires action at the European and even international level'.[11] Community action on the environment had always been defended in terms of the barriers to free trade which disparities in the stringency of environmental regulation can create. These barriers may be less apparent than, for example, differences between national regimes of indirect taxation on goods and services, but they are not necessarily less effective.

Subsidiarity is often presented as a necessary counterweight to 'qualified majority voting' which many see as a further erosion of national sovereignty: the British Cabinet and the Westminster Parliament could be united in their opposition to a measure which, having secured a qualified majority in the Council of Ministers, they are then obliged to implement. It must be remembered that QMV, like subsidiarity, was introduced not at Maastricht but in the Single European Act, signed on behalf of the United Kingdom by Margaret Thatcher in 1986. Adherents of her policies would no doubt observe that the principal purpose of this Act – the adoption of measures to assist the establishment of the internal market by 31 December 1993 – was in keeping with the orthodox aims of the original Treaty of Rome; it was only subsequently that it was heralded as a step on the heretical path to monetary and political union.

BOX 2.1 Article 130r, Treaty of Rome (as amended)

1 Community policy on the environment shall contribute to pursuit of the following objectives:

- preserving, protecting and improving the quality of the environment;
- protecting human health;
- prudent and rational utilisation of natural resources;
- promoting measures at international level to deal with regional or worldwide environmental problems.

2 Community policy on the environment shall aim at a high level of protection taking into account the diversity of situations in the various regions of the Community. It shall be based on the precautionary principle and on the principles that preventive action should be taken, that environmental damage should as a priority be rectified at source and that the polluter should pay. Environmental protection requirements must be integrated into the definition and implementation of other Community policies.

In this context, harmonisation measures answering these requirements shall include, where appropriate, a safeguard clause allowing Member States to take provisional measures for non-economic environmental reasons, subject to a Community inspection procedure.

3 In preparing its policy on the environment, the Community shall take account of:

- available scientific and technical data;
- environmental conditions in the various regions of the Community;
- the potential benefits and costs of action or lack of action;
- the economic and social development of the Community as a whole and the balanced development of its regions.

4 Within their respective spheres of competence, the Community and the Member States shall cooperate with third countries and with the competent international organisations. The arrangements for Community cooperation may be the subject of agreements between the Community and the third parties concerned, which shall be negotiated and concluded in accordance with Article 228.

Source: Article 130r(1–4) of Treaty of Rome 1957, as amended by the Single European Act signed in 1986 and by the Treaty on European Union signed at Maastricht on 7 February 1992.

At the risk of over-simplifying a complex issue, environmental legislation is now enacted under the authority of either Article 130s or 100a by a qualified majority of the Council of Ministers acting in cooperation with the European Parliament. Article 130s(2) lists three areas in which unanimity in the Council is required (and the Parliament's role reduced to that of consultation):

- provisions primarily of a fiscal nature
- measures concerning town and country planning, land use (with the exception of waste management), and the management of water resources
- measures significantly affecting a member state's choice between energy sources and the general structure of its energy supply

European case law and the environment

From the outset, the European Court of Justice (ECJ) has been required to interpret various provisions of the Treaty of Rome. These various rulings now amount to a considerable corpus of case law. For present purposes, attention will be concentrated upon a number of cases which have considered the fundamental duty:

> to take all appropriate measures . . . to ensure fulfilment of . . . obligations arising under Community law

imposed on member states by Article 5 of the Treaty of Rome, which set up the European Economic Community.

This duty encapsulates the loss of sovereignty which membership of the Community entails. It is of course a partial loss insofar as it impinges only upon those areas where the Treaty applies. Nevertheless, the use of such terms as 'all' and 'ensure' does little to assuage the concerns of those inimical to the European ideal, and encourages calls for subsidiarity and for a narrow interpretation of the Treaty. It is possible to cite four closely related principles of European Community law which, to greater or lesser extent, stem from this fundamental duty:

- supremacy
- direct effect
- sympathetic interpretation (or indirect effect)
- liability of a member state for damage resulting from its inadequate implementation of Community law

In this chapter, I seek to outline the relevance of the last three to contemporary environmental problems, in particular whether they can assist the enjoyment of rights conferred upon citizens of the Union by its environmental legislation. It is intended to establish a framework by which other specific examples – concerned with air pollution, water quality and wildlife conservation – appearing in subsequent chapters can be understood.

The relative recency of the Community's candour with regard to its environmental aspirations may explain why it is possible to examine the impact of case law on its environment programme, but it is not possible to point (save for perhaps one exception, see Box 2.2) to cases, arising out of recognisably 'green' issues, among those which have acquired a 'constitutional' significance. Of course, environmental cases – originating both in (Article 177) referrals from national courts[12] and from enforcement proceedings (Article 169) against defaulting member states – are beginning to add to European case law. When considering the consequences of developments in Community law in creating 'environmental rights' which could be recognised in English courts, it is necessary to observe:

1   that the majority of ECJ case law originated in member states with a written constitution and a legal system (e.g. Roman law, the Napoleonic Code) markedly different from that of the English common law tradition;

2   that these cases arose from disputes with little or no substantive environmental content.

Under s.3(2) of the European Communities Act 1972, courts in the UK are obliged to consider any relevant case law of the European Court. But notwithstanding this obligation, there remains a wide discretion which members of the English bench can use to resist what they might consider to be alien influences, inimical to legal certainty, if not to the proper development of English environmental law.

BOX 2.2 *Factortame*: an environmental or constitutional case?

Factortame[13] is undoubtedly a seminal case having a 'constitutional' significance, especially for the UK from where it originated. It further strengthened the doctrine of supremacy in that it established that domestic procedural law (the disability – Hohfeld's correlative of 'immunity' – of English courts to grant an interim injunction against the Crown) must be set aside if it impedes the realisation of Community law. Since it is concerned with access to a common property resource, *Factortame* has, by our definition, very real claims to be an 'environmental' case.

The principal issue in this case was the attempt by HM Government to use the Merchant Shipping Act 1988 to exclude certain foreign-owned (Spanish) but British-registered fishing vessels from gaining access to the

British fishing quota under the Common Fisheries Policy. Factortame Ltd was one of a number of companies which sought, via judicial review, to challenge the legality of those provisions of the 1988 Act which, by discriminating against boats which were not effectively in British ownership, denied them their rights under Community law (namely Articles 7, 52 and 221 of the Treaty). Fisheries are 'natural resources' and as such require 'prudent and rational utilisation' (Art. 130s(1) of the Treaty of Rome as amended) with catches limited by national quotas. The fact that the connection between the Common Fisheries Policy and environmental policy is so rarely made is perhaps eloquent testimony of the need for 'integration' of an environmental awareness in all Community policies.

Direct effect

In *Van Gend*, a case concerned with Article 12 (fiscal barriers to free trade) of the Treaty of Rome, the European Court held that Community law amounted to a

> new legal order [in which] independently of the legislation of Member States, Community law . . . not only imposes obligations on individuals but is also intended to confer upon them rights which become part of their legal heritage.[14]

Following *Van Gend*, case law of the European Court of Justice has elaborated the principle that under certain conditions Articles of the Treaty of Rome, although written in terms of duties imposed on member states, may implicitly confer rights upon individuals. *Van Duyn*[15] established that directives could similarly bestow rights upon individuals and that these rights could be capable of 'direct effect'.

Of the five forms of Community legislation set out in Article 189 of the Treaty, regulations are binding and form part of the law of each member state without further measures of implementation. In contrast, directives, although their objectives are binding, allow member states a certain measure of discretion in the manner in which they achieve those objectives. Where a directive has been inadequately implemented into domestic law or where compliance has been delayed, an individual may still be able to rely upon a 'directly effective' provision of that directive in the national courts. The status of 'direct effect' was strengthened with the *Ratti*[16] case, which introduced an argument:

> a Member State which has not adopted the implementing measures required by the directive in the prescribed periods may not rely, as against individuals, on its own failure to perform the obligation which the directive entails.

which is comparable with the English common law concept of 'estoppel' by which individuals may not gain advantage from unlawful acts (whether of omission or commission). Subsequent ECJ case law, in particular the ruling in *Becker*,[17] has elaborated the concept of rights originally articulated in *Van Gend* and has gradually come to describe the circumstances in which direct effect may be invoked:

> wherever the provisions of a directive appear, as far as their subject-matter is concerned, to be unconditional and sufficiently precise, those provisions may, in the absence of implementing measures adopted within the prescribed period, be relied upon . . . in so far as the provisions define rights which individuals are able to assert against the State.[18]

This would seem to have confirmed an important doctrine of Community law, leaving little room for uncertainty. In questioning this view, I examine the writings of Ludwig Krämer – a German lawyer and former Head of Legal Affairs in that part of the European Commission (Directorate General XI) most concerned with environmental protection – and those of Andrew Geddes (since 1994 an English judge). Inconsistencies between these two accounts cannot, I believe, be dismissed simply as differences in emphasis.

Direct effect of environmental directives according to Krämer

The first and most authoritative discussion of the implications of the doctrine of 'direct effect' for the environmental directives of the European Community was that of Ludwig Krämer[19]. His chosen examples were grouped under three headings, namely:

1   maximum values, maximum concentrations and limit values;
2   prohibitions;
3   obligations to act.

The first of these three categories – with its numerical values for maximum permitted concentrations of pollutants – is the least problematic. It is not incidental that the ECJ's most explicit recognition of rights – conferred upon individuals by an environmental directive – is to be found in a ruling concerned with limit values on (ambient) air quality standards for sulphur dioxide and particulates:

> whenever the exceeding of the limit values could endanger human health, the persons concerned must be in a position to rely upon mandatory rules in order to be able to assert their rights.[20]

However, it should be noted that this ruling was not concerned with direct effect; it was an Article 169 action brought by the Commission against Germany for its failure adequately to implement the directive in question[21]

30

Krämer's understanding of the 'direct effect' doctrine is that it is essentially a *sanction* to be used against a member state which has failed adequately to implement EC law. However, this particular instrument of correction was to be wielded by citizens; Krämer was well aware of the inability of the Commission (and the infringement procedures in the Treaty) to police the expanding environmental legislation effectively throughout the Community. His 1991 paper has recently been superseded by a more extended analysis[22] which takes account of case law in the intervening period. He argues that a provision of a directive has 'direct effect' if it satisfies five conditions:

1   the period for transposing the directive into national law must have expired;
2   the Member State has not or incorrectly transposed the provision into national law;
3   the provision must be unconditional;
4   the provision must be sufficiently precise; and
5   the provision must explicitly or implicitly confer rights upon individuals as against Member States.[23]

He also introduces the 'regulation test' (of the direct applicability of the provision of a directive): if the provision in question could be rewritten as a regulation which leaves no discretion to the member state and in which a right, conferred upon individuals, is clearly apparent, then that provision is capable of direct effect.[24]

It cannot be denied that the early case law from which direct effect emerged conceived it as a means of strengthening the rule of Community law ('l'effet utile'). A concentration upon its role as a sanction on defaulting member states can deflect attention away from the gradual shift in emphasis towards the individual citizen and to the positive rights conferred upon him by directives (and by certain articles of the Treaty).

At one point in his later account Krämer[25] suggests, from a consideration of *Becker*, that it might be more meaningful to speak of 'interests' rather than rights as the object of the implied protection of certain elements of Community law. Whether referred to as rights, interests or whatever, doubts as to their existence are ultimately resolved by the European Court of Justice. The cases heard hitherto, and on which the doctrine has been founded, have been concerned predominantly with issues, like discrimination and taxation, where the recognition of an individual right (or interests meriting protection) is relatively straightforward. However, this process of recognising rights or interests of individuals is, I suggest, more complex in regard to environmental directives.

Direct effect of environmental directives according to Geddes

In another, equally authoritative discussion of the direct effect of environmental directives, Geddes[26] makes the useful distinction between procedural rights and those which, in the absence of a more attractive term, I shall label 'substantive'. The former refers to the right to be consulted, to be informed, or to be involved in a decision-making process. The very purpose of 90/313/EEC is to establish a right of access to information[27]. In the same way, the *raison d'être* of Article 6.2 of 85/337/EEC[28] is to create the right of the public concerned to be involved (but this would not appear to extend to a right to be funded by legal aid; see Box 2.3) in the procedure by which development (with environmental impact) is approved.

BOX 2.3 Legal aid for third parties at planning inquiries

Participants at public inquiries in the UK receive no funding from the public purse. Where the inquiry involves a lengthy examination of complex arguments in favour of a development, individual objectors are put at a disadvantage compared with private developers or public authorities. Mrs Emily Sendall objected to the development of a clinical waste incinerator in Gateshead; she unsuccessfully sought legal aid to be represented at the public inquiry.[29] The application for judicial review of this decision was argued along the following lines. The environmental assessment of the incinerator was to be considered at this inquiry; Article 6.2 of Directive 85/337/EEC requires each member state to ensure 'that the public is given the opportunity to express an opinion before the project is initiated'. Her counsel argued that since this provision of the directive was precise, unconditional and implied a right of individuals, it had direct effect. In consequence the Legal Aid Board, as an emanation of the state, must respect the duty imposed by the Community provision which took precedence over any national restrictions on legal aid for public inquiries. Mr Justice Kennedy was unconvinced by the idea that Article 6.2 entailed obligations in respect of legal aid; as was the Master of the Rolls when dismissing Mrs Sendall's subsequent appeal.

Substantive rights, in contrast, are not created in such a clear and explicit way; they must be inferred from duties imposed upon member states to protect certain components of the environment. Geddes cites as exemplars directives which set limit values on harmful substances in the air or in drinking water. Both of these have a clear rationale in terms of protecting human health and were similarly cited by Krämer as examples in his first category (above). But the whole thrust of modern environmentalism is that we (humans) benefit in the long term from any measure taken to protect the environment. With some measures (for example, drinking water standards) the benefits are more immediate and obvious than others (such as restrictions on greenhouse gases). Attempting to identify environmental directives which are indifferent to human interests seems to invite confusion.

With a procedural right, like access to environmental information, there is little uncertainty as to the nature of that right (although there may be dispute over individual derogations over national security or commercial confidentiality). And while there may be disputes over the practicalities of presenting the information, there is no doubt as to the persons on whom this right is conferred, namely citizens of each member state. With substantive rights, it is sometimes necessary to infer (from the wording of the directive imposing a duty upon member states) both the benefits and the beneficiaries. This process of inference represents a further source of uncertainty. It may require a close reading of the text and preamble of an environmental directive in order to determine who exactly is being protected and from what threat.

Direct effect and locus standi

In order to rely upon a directly effective provision in a national court, a citizen must first gain access to that court. According to Geddes,

> There are many provisions of Community environmental law which have direct effect but where it cannot be said that those provisions were intended to protect health or welfare of individuals so as to create 'Community rights' which a national court must uphold.[30]

Geddes further argues that Directive 76/464/EEC,[31] limiting the discharge of toxic substances to inland and coastal waters, falls within this category. Irrespective of its precise and unconditional qualities, it does not create rights which individuals can rely upon in national courts in the event of incomplete implementation:

> the purpose of [76/464/EEC and its daughters] being to protect the aquatic environment and not human health, no *locus standi* to the individual to enforce these measures can be derived.[32]

This directive (along with its daughters) is one which Krämer also believed to be capable of direct effect. He does not elaborate the 'rights' implications but concentrates on the precise and unconditional nature of the limit values and quality objectives it lays down.[33] Geddes' assertion that a directive which does not protect the rights or interests of individuals can nevertheless be capable of direct effect cannot be reconciled with Krämer's analysis described above. There is something paradoxical about a directive which in theory is capable of direct effect but, because of a national law on standing, cannot be relied upon by *any* citizen of that member state.

By way of another example, Krämer cites the environmental impact assessment of an annex I project under 85/337/EEC in which all five conditions are met but the person invoking direct effect is not a member of the 'public concerned'.[34] This implies that it is possible for a directive provision to be successfully relied upon by one citizen of the defaulting member state but not by another; yet no guidance is given as to the qualifications for membership of the empowered group. Similarly, employees and residents near a major-accident hazard have a right, under Article 8 of the CIMAH directive[35] to be informed about evacuation procedures and other matters in the event of an accident. This provision is cited by Krämer[36] as a clear and precise 'obligation to act' which is capable of direct effect but, by implication, only by employees at that site or by residents in its vicinity.

Geddes' (and to a lesser extent Krämer's) approach points to two seemingly independent systems:

1   a category of articles of the Treaty and directives which, although expressed in terms of obligations on member states, implicitly confer rights (or interests) upon individuals. In certain circumstances (precise, unconditional, etc.) these rights can be relied upon by citizens in their national courts;

2   for each member state, a set of national laws governing *locus standi*, which is independent of Community law and which selects from that category (i.e. 1 above) of potentially directly effective Treaty and directive provisions, those which can actually be invoked in proceedings in national courts.

In *Rewe*[37] the ECJ established that national law, and not Community law, determines the remedies which apply in the event of infringement of Community rights. This independence is subject to two conditions: the remedy must be no less favourable than that which applies when actions in purely national law arise in comparable circumstances; and the national procedures must not make it impossible to exercise the right conferred by Community law. The ruling in *Rewe* relates to the remedies available after an infringement of Community law has been established; denial of standing is more serious since it precludes the possibility of a judicial hearing in which that infringement may be established. The applicants for a judicial review in *Twyford*[38] (see Chapter 3) were denied standing albeit the judge accepted that the directive provision, which they claimed had been infringed, was capable of direct effect. Geddes is not alone in criticising the judge's insistence upon applying a private law notion of standing in what was clearly an action in public law.[39] The case for the harmonisation of the rules of standing in environmental cases across the member states has been outlined by Rehbinder.[40]

Human or environmental interests?

The apparent conflict between two authoritative commentators (Krämer and Geddes) is hardly surprising once the labyrinthine nature of the relationship between human interests and the environment is recognized. As our understanding of the complexities of environmental systems develops, so the notion of (substantive) rights implicit in certain directives and of traditional rules on standing seem equally irrelevant to contemporary challenges.

For Geddes, it is a matter of textual analysis whether a measure designed to protect a part of the physical environment (the oceans, the atmosphere or the habitat of an endangered species) also serves to protect or promote tangibly human interests. If the title or the preamble of a directive includes a phrase like 'drinking water', then the process of inference need not be protracted. However, if as with the Directive on dangerous substance in the aquatic environment, the threat to human health is less direct and dependent upon certain assumptions concerning dispersion and persistence of the pollutants, accumulation in food chains, etc., then the attribution of rights (and hence direct effect) is withheld. By way of contrast, Geddes implies[41] that two directives,[42] concerned *inter alia* with the quality of the waters from which shellfish for human consumption are caught, are sufficiently related to health to be directly effective.

Inferring individual rights or interests (and hence certain opportunities to redress faulty implementation) from those environmental directives which have a subordinate aim of protecting health in a clinical sense is, I suggest, outdated. Modern environmentalism stresses the fatuity of searching for aspects of the physical environment which do not impinge, to some finite extent, upon human interests in the widest sense. What of directives whose primary purpose is the protection of non-human species?

As an example of his second category – prohibitions – of directly effective directives, Krämer cites Directive 83/129/EEC,[43] which requires member states to take measures to prevent the commercial importation of fur skins and other products derived from two species of seal. Public condemnation of the annual hunting of young seals in Canada and Norway resulted in the European Parliament calling for legislation. Krämer also implies that Articles 5–9 of the Birds Directive[44] (like the Seals Directive 83/129/EEC) are sufficiently precise and unconditional to be capable of direct effect, but he makes no specific reference to rights or interests at this point.[45] Any rights or interests might be interpreted as being conferred upon the seals or protected bird species; but, as Krämer makes clear in a later paper, Community law does not recognise the rights of non-human species.[46] Therefore, if the protection of the rights or interests of human individuals is a *sine qua non* of direct effect, then the Birds Directive (and its cognates) cannot have this property.[47] This conclusion can be avoided if one accepts that the protection of animal and plant species can be interpreted, without too much distortion, as a cause which benefits humans: many of Europe's citizens feel deeply saddened at the loss of the habitats of threatened species; others derive some satisfaction that the cruelty associated with the culling of certain species of seal does not occur on Europe's northern shores (outside Norway).

The Community was set up so that all its citizens could enjoy the benefits of free trade within a common market, but only someone (an importer, for example) who has actually suffered as a result of a tariff or trade barrier contrary to a directive (or Treaty provision) could invoke direct effect in a domestic court. Thus the environment is not the sole source of these public *v.* private conflicts even if it does produce some of the most interesting examples. The House of Lords Select Committee has advocated allowing 'citizen suits' as a means of encouraging compliance with Community environmental law.[48] If a proposal, not confined to environmental law, of the Law Commission becomes law, the High Court will be empowered to hear judicial reviews brought in the public interest by individuals with no further personal concern.[49] Even in advance of this proposal, it could well be that English judicial attitudes to standing are already more relaxed than the European equivalent.

Three residents of Tahiti sought to challenge a decision of the Commission not to interfere with the French nuclear tests at Muroroa (in the Pacific Ocean). Article 147 of the Euratom Treaty (almost identical in wording to Article 173 of the Treaty of Rome) allows individuals to challenge the legality of actions by the Commission and other institutions if they can show 'direct and individual concern' to themselves. The Court of First Instance, following previous case law, held that the action in question must affect the applicants in a way which can be distinguished from that on all other persons. If the three applicants were posed some increased radiological risk from the nuclear test, then all other residents of the region were similarly affected; hence the application fell.[50]

A strict interpretation of 'direct and individual concern' effectively rules out Article 173 actions in truly 'global' environmental problems. Regulation 594/91/EEC bans the use of certain compounds which deplete the ozone layer and thereby increase the risk of skin cancer. If the Council were to rescind or, in some other way, modify this or any similar regulation on ozone depleting chemicals, it is difficult to see how any individual or environmental group could claim the 'individual concern' necessary to challenge the legality of that action under Article 173. The 'lead in petrol' issue of the late 1970s suggests another, less hypothetical example.

The Commission's decision to include (under pressure from the United Kingdom[51]) a mandatory *lower* limit of 0.15g/l in Directive 78/611/EEC frustrated the desire of certain member states, Germany in particular, to proceed rapidly to exclusive reliance upon leadfree petrol. Had the plaintiffs in *Budden* (see Chapter 1) chosen to pursue their objectives not via the English civil courts, but by invoking Article 173 to challenge that decision, they too presumably would have fallen foul of the 'individual concern' criterion unless, as children living near an urban motorway, they were adjudged to pass that test.

Francovich *liability*

The *Francovich*[52] ruling extends and refines the principle[53] that national courts must ensure real and effective protection for Community-derived rights:

> where . . . a Member State fails to fulfil an obligation imposed upon it by Article 189(3) of the Treaty to take all the necessary steps to achieve the

result required by the Directive, that provision of Community law, to be fully effective, must give rise to liability for damages provided that three conditions are fulfilled:

1   The result required by the Directive must include the conferring of rights for the benefit of individuals;
2   The content of these rights must be determinable by reference to the provisions of the Directive; and
3   There must be a causal link between the breach of the obligation of the state and the damage suffered by the person affected.[54]

Francovich arose from Italy's failure in regard to a directive on protection of employees against employer insolvency. Its potential impact on Community environment law has been the subject of inconclusive speculation in the literature.[55] Lord Slynn, a former Advocate General, claims that 'the *Francovich* decision may have a significant effect in the application of environmental rules'.[56] In theory a directive, with a discretionary element sufficient to fail the test for direct effect, may nevertheless be the subject of an action for damages against a defaulting member state. Remarks to that effect were stated *obiter* in a ruling by a Dutch court after 76/464/EEC (Toxic Substances in Water Directive) was adjudged (contrary to both Geddes and Krämer) not to have direct effect.[57] Irrespective of the absence of the 'precise and unconditional' criteria, I shall argue in subsequent chapters that there remain considerable obstacles in the way of action in *Francovich* to promote the rights 'to breathe clean air' and 'to bathe in clean seawater' in the UK.

Francovich liability arises when a directive has been inadequately implemented. More recently, the ECJ has extended the notion of state liability to include compensation to individuals harmed as a result of breaches of the Treaty itself. The latest ruling[58] in the aftermath of *Factortame* is concerned with measures taken by a national legislature which violate an article of the Treaty. Two of the conditions – the rule of law infringed must have been intended to confer rights on individuals; and a causal link between the infringement and the harm sustained must be established – are retained from *Francovich*. The third is novel and depends upon the degree of 'discretion' which the member state enjoys in its legislative function. Where a wide margin of discretion exists, liability should be confined to the most serious violations of Community law. Conversely, where little or no discretion is allowed, liability arises, and compensation is payable, when instances of lesser harm are suffered by individuals.

Writing before the ruling in *Brasserie* by ECJ, Coppel invokes a floodgates vision in which 'astronomical' amounts of damages would undermine the ability of governments to govern if they were to become liable for 'every *prima facie* breach of Community law by a Member State which occasioned harm to any individual'.[59] He presumably therefore would welcome the latest development. Schockweiler, similarly concerned, argues that the term 'right' (as it appears in the three conditions for Francovich liability) should be interpreted so as to exclude 'mere interests'.[60] Krämer, as noted above, believes that both rights and interests can, provided the other criteria are met, similarly be protected by direct effect. Any formal distinction between rights and interests would have to be made by the ECJ, and since that body gave birth to the doctrines of 'direct effect' and state liability, this would seem unlikely.

Indirect effect

Writing in early 1996, Jane Holder reports that '[t]he current success rate of individuals seeking to rely on provisions of Community environmental law via the direct effect doctrine is nil'.[61] No indication is given of the rigour of the search for such examples in all relevant jurisdictions within the Community, but it is possible now to point to one positive ruling by the ECJ (*Aanemersberdijf*; see below), but even then the language of direct effect was avoided. The barriers to the use of direct effect would seem considerable; these may be practical (lack of funding, information, etc.) but meeting the precise, unconditional and vertical criteria is clearly not trivial. This observation has led some commentators to speculate upon *Francovich* as an easier remedy for inadequate implementation of environmental directives. Less attention has been given to potential of indirect effect.

The doctrine of 'indirect effect' (or sympathetic interpretation) has a shorter history than direct effect, and to a large extent it represents a reaction to the European Court's opposition to horizontal direct effect (namely an individual relying upon a right against another individual rather than the state). The decision in *Marleasing*[62] was again founded upon the (Article 5) duty on member states to take all appropriate measures to assist the Community's aims:

> It follows that in applying national law, whether the provisions concerned pre-date or post-date the directive, the national court asked to interpret national law is bound to do so in every way possible in the light of the wording and the purpose of the directive.[63]

For a provision of Community law to have direct effect, it must be 'justiciable', which Pescatore takes to mean 'capable of judicial adjudication, account being taken both of its legal characteristics and of the ascertainment of the facts on which the application of each particular rule has to rely'.[64] This view would seem to apply no less to indirect effect. Indeed indirect effect would seem to encompass a greater realm of justiciability – one which is wider than (but embracing) those provisions of directives which pass the special test of precise, unconditional, etc.

Indirect effect is not without its own obstacles. If a directive is not worded in precise and unconditional terms, it allows more scope for a restrictive interpretation of a member state's obligation. Some judges might argue that the principle of subsidiarity required them to use that discretion in favour of the member state. Few successes[65] will be described; but equally, none of the reported failures has really tested the utility of this doctrine as a tactical device for environmental activists.

The *Duddridge*[66] case is, in many respects, a poor example to cite: there is no clear distinction between direct and indirect effects; the applicants are relying upon part of an article of the Treaty rather than the provision of a directive. Since the Article in question is 130r of the Treaty (as amended) and in particular 'the precautionary principle' (see Box 2.1 above), it is necessary to give it some attention, not least in order to consider the prospects of substantive 'rights' emerging from other principles of Community environmental law (see Case Study below).

In another 'bad example', the ECJ, after ruling that the directive on cadmium discharges to the aquatic environment is not capable of direct effect, would appear to have articulated a limit to the application of indirect effect. The wider relevance of this case (discussed further in Chapter 5) is open to question: the Italian state, notwithstanding the thirteen years in which it had failed adequately to implement the directive, was relying upon both doctrines to take criminal action against a company discharging without an authorisation. The Court refused to interpret the doctrine as imposing upon an individual an obligation (and a criminal liability in particular) which, although contained in a directive, had not been transposed into national law.[67]

Wychavon[68] represents another case in which a negative ruling on indirect effect followed rejection of direct effect. The reverse order might seem more logical, but it is necessary to reiterate that indirect effect arose after (and as a reaction to the exclusion of horizontal) direct effect. Fitzpatrick's analysis[69] demonstrates the flawed nature of the ruling in this case; and little point is served by considering it here, were it not for the fact that it was cited as authority in yet another judicial review[70] involving the direct and indirect effects of an article of the Environmental Assessment Directive. When it comes to hearing the appeal granted to the applicant in this latter case, the Court of Appeal, unlike any court which has previously considered 85/337/EEC, will have the benefit of the definitive ECJ decision in *Aanemersberdijf*.[71]

Here the European Court ruled, in a referral from the Netherlands State Council, that modifications to certain dykes should not be allowed to proceed in the absence of an assessment of their resulting impact on the environment. After arguing that dykes did indeed fall within Annex II projects (namely those for which some size threshold must be exceeded before an assessment is obligatory) the ECJ then navigated previously uncharted waters by arguing that a member state's discretion to set thresholds could not be used so as to exclude whole classes of project, such as dyke modification, from the scope of the directive. Conventional understanding (including Krämer's) held that the discretion, given to member states in setting thresholds for Annex II projects, meant that Article 4.2 failed the test for direct effect. This point was made by the Dutch and UK governments when intervening in this case. When supporting the views of the applicants in the original action, the ECJ avoided the language used in earlier cases on direct effect. Instead it chose a terminology closer to sympathetic interpretation and focused upon the obligation of a national court to consider, and if necessary to declare as illegal, any use of discretion which subverted the underlying aims of the directive.

No doubt the applicants in this Dutch case were indifferent as to whether it was direct or indirect effect which ultimately secured their success. Notwithstanding the number of comparable disputes over the need for environmental assessment, no similarly successful outcome can be found within the UK. Some of these disputes will be considered further in the next chapter; I will simply note here that no court (or planning appeal) in the UK has yet chosen to refer to Luxembourg for a ruling on this most vexatious directive.

CASE STUDY
The precautionary principle and English law

A judicial review of regulations, made by the Secretary of State for Trade and Industry, offered a group of residents concerned at the effects of nearby high-voltage electric cables[72] an opportunity to exploit Community law remedies in a case where traditional, domestic procedures seemed to hold little promise. It involved both environment and health: electromagnetic fields in the vicinity of high-voltage power cables were alleged to give rise to an enhanced risk of leukaemia. Many of the issues on the role of epidemiological evidence in establishing causation are discussed (in relation to action in tort) at more length in Chapter 6; but before examining the legal arguments, it is necessary to describe 'the precautionary principle'.

Vorsorgeprinzip appears in the English version of the Maastricht Treaty as the 'precautionary principle', but it defies precise translation. According to one recent interpretation, it embraces six basic concepts:[73]

1 preventative anticipation
2 safeguarding of ecological space
3 proportionality of response or cost-effectiveness of margins of error
4 duty of care or onus of proof on those who propose change
5 promoting the cause of intrinsic natural rights
6 paying for past ecological debt

Some of these concepts – cost effectiveness is the obvious example – are deeply rooted within the English tradition. Placing an onus of proof on those who propose change represents a departure from a landowner's privilege to do as he wishes provided the effects are confined within his boundaries; but, following half a century of controls over land use, it does not amount to a radical change. Others – safeguarding of ecological space –

stem from recent recognition of the fragility of natural ecosystems and of an awareness of the uncertainties in contemporary scientific understanding of the long-term effects of man's impact upon them. The notion of 'intrinsic natural rights' of ecosystems and of non-human species originates in a distinctly *ecocentric* moral viewpoint and could actually conflict with purely anthropocentric constructs like proportionality and cost-effectiveness. However, discussion in *Duddridge* was to be focused on the first four (anthropocentric) concepts.

The hypothesis that environmental exposure to (high-frequency, non-ionising) electromagnetic radiation arising from electricity transmission lines could be associated with an increased risk of cancer dates from the late 1970s. Table 2.1 summarises the findings of an early American study from which the relative risk (i.e. the ratio of the risk of exposed children over non-exposed dying of cancer) of 2.2 may be estimated. Other studies have failed to show an association. The National Radiological Protection Board is the UK body with a statutory duty to advise the Secretary of State on such matters. The Board set up an Advisory Group, under the chairmanship of the eminent epidemiologist, Sir Richard Doll, to consider the latest evidence. It concluded in 1994:

> The studies do not establish that exposure to electromagnetic fields is a cause of cancer but, taken together, they do provide some evidence to suggest that the possibility exists in the case of childhood leukaemia. The number of affected children is however small.
>
> Experimental studies to date have failed to establish any biologically plausible mechanism whereby carcinogenic processes can be influenced by exposure to the low levels of [electromagnetic fields] to which the majority of people are exposed.[74]

Table 2.1 Electromagnetic fields: case-control study

	Exposed	*Not exposed*
	Residence at death close to source of high electromagnetic radiation	*Residence not close to source of high electromagnetic radiation*
Cases: children dying of cancer in Colorado 1950–73	129	199
Controls: living children matched for month of birth	74	254

Source: N. Wertheimer and E. Leeper, 'Electrical wiring configurations and childhood cancer', *American Journal of Epidemiology* (1979) 109, 273.

This statement was submitted in evidence at a judicial review of a decision of the Secretary of State for Trade and Industry to decline to issue regulations to (principally) the National Grid Company under the Electricity Act 1989 to restrict electromagnetic fields from power cables. The challenge was brought by the parents of three children, resident in an area of northeast London where the National Grid Company was then laying a new high-voltage underground cable. Even though a causal connection was far from proven, the evidence of a possible risk was – in the light of the precautionary principle – sufficient to necessitate the inclusion of preventive measures in regulations.

There was little dispute between the expert witnesses appearing for each side. As a corollary to the statement quoted above, the Advisory Group argued for statistically robust studies based upon objective measurements of exposure. The two sides differed on the legal implications for the Secretary of State on this 'possibility' of a causal connection.

There was only one legal point at issue: whether Community law obliged the Secretary of State to apply the precautionary principle. If so regulations should require the National Grid Company to ensure that fields from their cables do not exceed 0.2 micro-tesla at the nearest point of any residential property. The subordinate question – whether the 'possibility of harm' was sufficient to justify the principle's application in this case, were it to be an obligation – was soon answered by Smith, J.'s acceptance that it was.

Mrs Justice Smith noted that there was no authoritative definition of the precautionary principle. UK commitment to it was stated in the 1990 White Paper. It had also been cited in an Australian case, in which the judge declared it to be 'a statement of common sense'. Junior counsel for the applicants happened to be the co-editor of the book *Interpreting the Precautionary Principle*, from which the sixfold structure quoted above was taken. Another extract from this book was read in Court:

> Now that the Maastricht Treaty is ratified the precautionary principle will apply to a British Minister when, as a member of the Council, he contributes to the formulation of EC policy by agreeing the form of words in an item of EC legislation. The principle applies to Community policy and does not apply to any aspects of purely national policy which are not part of EC policy.[75]

But the author of this chapter (Nigel Haigh) was not a lawyer and the book carried 'no great authority'. Smith, J. came to the view that the principle was intended to protect the environment itself and was not concerned with damage to human health attributable to environmental factors.[76]

The absence of a clear definition was not to prove a hindrance to an unequivocal ruling: Article 130r was a statement of the environmental *policy* of the Community, and a binding obligation of Community law could not (in advance of any specific directive, decision or regulation) derive from 'a statement of policy, still less a statement of principles to underlie a future policy'. This reasoning was upheld by the Court of Appeal on 5 October 1995 when refusing leave to appeal. In reiterating the distinction between policy and law, this case becomes another in that recent spate of judicial reviews which involve an incantation of the domestic interpretation of the doctrine of separation of powers.

From its wording, Article 130r(1–3) (see Box 2.1) is clearly meant to be a statement on the structural components of future environmental policy. One might be forced to conclude that overall it can never meet the 'precise' test for direct effect. Thus environmental NGOs are unlikely to be able to challenge a member state deemed to have breached its obligation to observe the principles contained in this Article. However, some of these principles (precaution, high level of protection, integration) are more beset with imprecision than others. It is possible to conceive of a legitimate challenge by a responsible environmental NGO against a ministerial authorisation which did not demand rectification at source[77] or against a programme which paid subsidies to companies to desist from polluting activities. A vernacular understanding of terms like 'source' and 'polluter' may be sufficient to be justiciable without recourse to some technical references of questionable authority.

Mrs Justice Smith's reasoning in *Duddridge* is undoubtedly cogent, but if it were to be applicable to all the principles, not simply precaution, it does condemn Article 130r – the core of the Community's environmental law – to permanent exclusion from the 'first division' of Treaty articles (headed by Article 12, Fiscal Barriers to Free Trade) which can be invoked by individual citizens. Does the word 'policy' adequately describe the commitment to the 'balanced and sustainable development of economic activities' which now appears in the very aims of the Union (Article 2)?

Conclusions

With the wisdom of hindsight, Krämer's expectations of the direct effect of environmental directives now seem unduly optimistic. Why is it not possible to cite more instances of the successful exercise of Community-derived environmental rights?

Anyone anxious to invoke direct effect (indirect effect or Francovich) must first have the opportunity to appear in some national court or tribunal before he can persuade a judge that he has been denied some benefit intended in the particular provision of a directive, regulation or Treaty article. The dearth of such occasions

– or at least of those which have been reported – cannot be explained simply by confusion over standing, even if that were to be rife. Anterior to the issue of standing is the occurrence which leads to some perceived grievance, to a dispute which is potentially justiciable. The greater the rate of such occurrences, the more frequent recourse to litigation, and the greater the likelihood of seminal case law.

Van Gend was an importer of chemicals (unlawfully taxed); Von Colson's grievance involved equal opportunities; Francovich was an employee whose salary was not adequately safeguarded; Factortame was a company concerned about a barrier to trade. *Marshall*[78] represents one of the more important 'constitutional' cases within the case law of the ECJ. Although it failed to extend the 'direct effect' of directives from the vertical to the horizontal plane, it established that a body like an area health authority amounted to an 'emanation of the state'. It originated in an industrial tribunal where Miss Marshall, challenging her dismissal on grounds of sexual discrimination, successfully invoked the 'direct effect' of Article 5.1 of the Equal Treatment Directive.[79] Following an appeal by the employers, the Court of Appeal referred the matter to the ECJ (under Article 177 of the Treaty) who confirmed the Tribunal's original decision. Of the nearly 22,000 cases decided by industrial tribunals in 1995, some 704 were concerned with sexual discrimination.[80]

Most of the public law actions with an environmental and European dimension, cited in this book, have involved an appeal against refusal of planning consent. Reasons why the majority of tangibly environmental disputes should originate from the 10,000 appeals arising annually in England and Wales are presented in the following chapter. The great majority of planning appeals will – like the planning applications from which they derive – be concerned with small-scale matters like garages, house extensions, advertisements, etc. Nevertheless, reasons why discrimination cases have proven to be a more fruitful source of case law cannot be founded upon relative numbers.

Motorway extensions in areas of outstanding natural beauty, mineral development in national parks, damage to sites of special scientific interest: each one of these may occasion outrage and each one could generate important judicial precedents; but they total less than a hundred per annum. However, it is possible to cite other forms of environmental 'occurrence' which, being less concerned with procedural rights and more with personal health, can generate something closer to that private and personal grievance which victims of sexual discrimination perceive. Chapter 5 presents data from which it follows that 'occurrences' – of individuals drinking water or swimming in coastal waters which fall below Community prescribed standards – number hundreds of thousands annually in the UK alone. Even if one ignores the large numbers of victims in each instance of a defaulting supply zone or bathing beach, the reported rates of non-compliance – and in turn the potential for litigation – remain daunting. Drinking and bathing water were involved in the two instances in which the ECJ has reprimanded the UK government on its environmental record. The government's tardy response to the drinking water prosecution prompted Friends of the Earth to take a judicial review in which 'rights' were a central issue. Despite some compelling arguments in the Court of Appeal, the applicants were ultimately unsuccessful, notwithstanding the acceptance of direct effect.

If lack of opportunity cannot account for the dearth of 'successes' by environmentalists, what can? The failed attempts, described in following chapters, are now sufficiently numerous to allow a generalisation to be made. There would appear to be a reluctance among some English judges to accept the relevance, to environmental disputes, of a framework of Community remedies created by the European Court in cases concerned with other (i.e. non-environmental) rights, which have enjoyed a longer and less problematic protection by the Treaty of Rome. The problem is that this framework has yet to acquire a robustness sufficient to compel the more sceptical members of any judiciary (or executive for that matter) to accept the full implications of the Article 5 duty to 'take all appropriate measures' to fulfil Community policies.

3

TOWN AND COUNTRY PLANNING

Introduction

The history of town and country planning has been described as one in which landowners' rights to develop their land as they might wish have been progressively 'nationalised'.[1] And since planning entails one of the most obvious invasions of the rights of property owners, there are strong reasons for deferring its discussion until Chapter 7. However, some idea of the structure of the UK planning system is necessary in order to understand many of the issues which are raised in preceding chapters. Coupled with this vast erosion of proprietorial rights there are a large number of powers exercised by local planning authorities in the interests of their inhabitants. In addition, there are various procedural rights which private individuals can invoke in order to influence the planning process. Much of town planning's ever-expanding case law has originated in appeals brought by third parties. Public inquiries, associated with appeals against refusal of planning consent, have often become the forum in which more substantive issues have been debated.

Although never intended for that purpose, the public inquiries into various nuclear installations (see Chapter 6) gave environmental groups opportunities to subject UK energy policy to a scrutiny far more searching than anything that the parliamentary opposition could achieve. Planning has a more obvious part to play in regard to preservation of wildlife habitats; Chapter 8 describes the problems associated with the virtual exclusion of agricultural land from the planning process. Planning has always been concerned with the preservation of landscape; therefore an attempt to invoke 'direct effect' to prevent (highways) planning procedures from allowing an extension of the M3 motorway to destroy Twyford Down merits attention. It is also possible to point to occasions in which planning inquiries have been used as forums for the pursuit of such substantive aims as clean air. This chapter will describe one example where, perhaps for the first time, the arguments of a local residents' group were based upon case law of the European Court.

The intensity of the rhetoric associated with deregulation, the encouragement of the enterprise culture and the hostility which existed between central and local government, invite the question as to why a more extensive dismantling of planning systems was not attempted in recent years. In the UK a whole tier (namely metropolitan county councils) of government was abolished in the major conurbations; utilities, in public ownership for four decades, have been privatised. The right to develop land in one's ownership – which might seem a prime target for those most ideologically committed to a 'minimal state' – has yet to witness substantial denationalisation. In fact, the town and country planning system has been described as having 'in many ways . . . served the country well and the [Thatcher] Government has no intention of abolishing it'.[2] One of those many ways must include its value in assisting compliance with certain obligations of the European environmental programme. The UK's long-running antipathy to environmental assessment was frequently accompanied by declarations that it was unnecessary since the town planning system already incorporated such safeguards. When this opposition was finally overcome, the planning system[3] was considered to be the appropriate mechanism for implementing most of the requirements of the Directive. A number of instances in which environmental groups have attempted to invoke 'direct effect' of 85/337/EEC, after claiming its implementation to have been defective, are discussed below. Additional controls over hazardous substances have been grafted on to the planning authorities' remit; and contaminated land (Chapter 7) is but the most recent addition to the environmental duties of local authorities.

Town and country planning is implicitly environmental in the sense that its *raison d'être* is to regulate our immediate surroundings, allowing for a rational allocation of land uses among competing demands. It can also influence 'hard' issues like air quality, water pollution and the siting of hazardous installations which attract a narrower sense of the term. This influence, which is not confined to development which requires a formal environmental assessment, has sometimes been a source of conflict with other regulatory agencies, prompting ministerial advice on the need to respect boundaries. The tone of the admonition tended to vary and, most recently, the UK government departments[4] principally concerned with the environment have been anxious to stress the positive role that land use planning can play in the pursuit of sustainability, especially in reducing both the need for travel and reliance upon the private motor car.

Individual rights and land use planning

Rights, especially standing to seek judicial review of ministerial decisions, may assist 'open government', but their praises are most often sung by those who advocate less government. In view of the exaltation of individualism in recent years, it is hard to recall the hopes which were invested in the early 1970s in what were then the 'new-style' development plans. The need for 'public participation' was stressed in the Skeffington Report[5] and the involvement of a wide range of community groups and what are now known as NGOs (non-governmental organisations) in the preparation, under the benign direction of subsequently (1974) reorganised county planning authorities, of structure plans would enable the planning system to become 'the pace-setter in a more open and participatory style of government'.[6] But by 1978 (a year before Margaret Thatcher's first premiership), Professor McAuslan was forced to conclude that:

> while the rhetoric was that of participation, the practice remained that of public interest and private property.[7]

When structure plans were first introduced in 1968, the Secretary of State had to give each objector an opportunity of being heard at the 'examination in public'; this requirement was abolished in 1972.[8] Participation by the public or, more curiously, by any planning authority at the examination in public of a unitary plan[9] or of the modification of a structure plan[10] is now at the invitation of the Secretary of State. Individual participation as of right could be open to abuse, and the proceedings effectively nullified, by any pressure group whose members attended en masse but with each demanding to make an individual objection. However, it is reported[11] that a development plan generates on average 1,200 objections, and some ninety inquiries into development plans were anticipated to open in 1996–7. Save for the right of 'any person aggrieved' to apply for statutory review (number 7 in Box 3.1 which summarises the rights which third parties currently enjoy under the general planning system), there is no reference to rights associated with development plans. In practice, the position of 'third parties' is not as bleak as Box 3.1 suggests. In 1973 it was held[12] that a person, invited by a planning inspector to appear and make representations at an inquiry, becomes a 'person aggrieved' and thereby acquires the standing to seek a statutory review if so minded. This liberal approach in regard to planning preceded the reform of the procedures by the Supreme Court Act 1981 and, more to the point, the recent relaxation of judicial attitudes to the 'sufficient interest' criterion.[13]

BOX 3.1 Rights of 'third parties' under planning law in England and Wales

Information

1 Any member of the public has a right of access to the register of planning applications at all reasonable hours s.69(5).[*]
2 For certain classes of development (designated in the General Development Order and including those associated with an environmental assessment) the applicants are required to advertise their proposal in a local newspaper s.65. Any person then has twenty-one days in which to make representations to the planning authority s.71(1).

3 Applicants also have a duty to notify owners (tenants and tenants of agricultural land) of any land to which the application related s.66, who then have twenty-one days in which to make representations to the planning authority s.71(2).

Appeal

4 Only the applicant has the right of appeal against refusal of planning consent or against a condition applied s.78. There is no right of appeal against planning approval.

5 At an appeal involving a public local inquiry, the appellant, and any person who made representations under s.66 have a right to put their case before the inspector and to cross-examine the witnesses of the other parties.

6 Third parties, including those who made representations under s.71, have no right to be heard, but the inspector is empowered to permit their participation. Any member of the public has the right to inspect a 'statement of case' submitted by any of the parties, Town and Country Planning (Inquiries Procedure) Rules 1992 (SI 1992, no. 2038) r.6(9).

Statutory review

7 Any person aggrieved by a development (unitary, local or structure) plan may question its validity by applying to the High Court s.287.

8 Any person aggrieved by an order, direction or decision (especially an appeal) of the Secretary of State may question its validity by applying to the High Court s.288.

* Except where otherwise denoted, references are to sections of the Town and Country Planning Act 1990.

Although some recent judicial reviews by environmental pressure groups have received a great deal of publicity, it is necessary to recall that this acceptance of a greater role of 'third parties' owes much to the efforts of numerous local amenity societies, architectural conservation groups and other representatives of the 'soft' end of the environmental spectrum at countless planning inquiries. Some idea of current numbers of planning appeals and the methods of their determination is given in Table 3.1.

Table 3.1 Planning and enforcement appeals, England, 1995–6

	Planning appeals	*Enforcement appeals*
Determined by:		
Inspector	11,038	1,824
Secretary of State	176	62
Total	11,214	1,886
Percentage determined by:		
written submissions	82%	62%
hearing	11%	7%
local inquiry	7%	31%

Source: The Planning Inspectorate Executive Agency, *Annual Report and Accounts for the Year Ended 31 March 1996* (1996, HMSO) para 3.7.

In general, development now requires the prior consent of the local planning authority; appeal is to the Secretary of State (whose decision may be challenged by statutory review) but, since no right has been infringed, refusal does not normally entail compensation. In fulfilling its statutory duties, a planning authority seeks to mediate conflicts over allocation of land use so as to maximise the overall interests of the area. This task is unquestionably utilitarian in character; it is accepted that some interests may benefit whilst others may suffer (even to the point of encroachment upon their rights; see Box 3.2). The 'losers' do not thereby acquire a right of action in negligence. Case law has established that planning law does not impose on planning authorities a duty of care to third parties.[14] Equally, the existence of a planning consent for development which gives rise to nuisance does not automatically create a defence in any subsequent action;[15] such extinction of a private right without redress or compensation would be contrary to the established principles of English law.

Development is given a very broad definition: 'the carrying out of building, engineering, mining or other operations in, on, over or under land, or the making of any material change in the use of any building or other land'.[16] In principle therefore, the construction of the smallest garden shed, the erection of a lamp post or telephone box, and the conversion of a backstreet shop to an office all require planning permission. In order that the system is not overwhelmed by such trivial items, two important statutory instruments[17] grant planning permission for large classes of operations and a number of changes of use of land and buildings.[18] When deciding whether to grant consent for development which falls outside the various 'permitted' categories, a local planning authority must have regard to any relevant development plan and to any other material consideration.

BOX 3.2 Planning v. the right to one's home

Mrs Buckley is a Gypsy; shortly before the birth of her third child in 1988, she abandoned her itinerant life and moved her three caravans onto a plot of land which she owned in the district of South Cambridgeshire. Planning permission for caravans on this site was refused in 1989, and the Council began enforcement proceedings in 1990. Mrs Buckley's appeal against the notice was unsuccessful. The Inspector was conscious that the site posed road safety hazards and that it intruded into open countryside, contrary to the aim of the structure plan. These and other objections outweighed the need for more authorised sites for gypsy accommodation; and the Secretary of State accepted the Inspector's recommendation to dismiss the appeal.

In a subsequent application to the European Commission of Human Rights, Mrs Buckley argued that this enforcement action – preventing her from living in caravans on her own land with her family – contravened Article 8 of the Convention (reproduced in Box 1.3). By a majority of seven votes to five, the Commission held that a violation had arisen and referred the case to the Court.

(A related allegation that the criminalisation of unauthorised camping discriminated against Gypsies, in violation of Article 14, need not be discussed here.)

The UK government did not dispute that the enforcement action could amount to 'interference by a public authority' (Article 8.2) with the applicant's right; but this was 'in accordance with the law' and in pursuit of legitimate aims, namely 'public safety, the economic well-being of the country, the protection of health and the protection of the rights of others'. After considering the report of the Inspector at some length and the Secretary of State's reasoning in accepting the Inspector's recommendation, the Court came to the view that the UK authorities, in weighing the policy considerations against the applicant's right to respect for her home, reached their decision without exceeding their margin of appreciation. The Court did not see it as part of its task 'to sit in appeal on the merits of that decision', and by six votes to three found that no violation of Article 8 had occurred.

Had the Court found otherwise, the implications for the UK town and country planning system would have been incalculable. Coming from a body committed to human rights, the finding in this case serves as a particularly cogent reminder that owners of land have no automatic right to use their land – even for the creation of a family home – as they might wish.

Source: *Buckley v. the United Kingdom* (1996) JPL 1018.

The development plan is a mechanism by which the land use implications of various social and economic policies within a defined locality are addressed. Structure plans for counties and unitary development plans for the unitary authorities are strategic in character, comprising broad-brush attempts to reconcile a range of conflicting demands; whilst a local plan tends to be more prescriptive, identifying preferred options (housing, recreation, open space, green-belt, commercial) at all locations within its area. There is an obvious sense in which a development plan is 'environmental' in that it influences the allocation of *all* land uses within its area. There are certain land uses which are, in vernacular terminology, more 'environmental' than others. At the one extreme (soft) are the national parks and areas of outstanding natural beauty (see Chapter 8), the boundaries of which must be delineated and protected for one clearly recognised set of reasons; and at the other (hard) extreme, there are special industrial areas, nuclear installations and potential sources of major chemical hazards which must be identified and accorded particular attention for very different reasons. The term 'environmental' is applied both to the motivation of those structure plan policies which seek to preserve those qualities which give the Lake District its distinctive character, and to the rationale for policies which underpin a presumption against further residential development in the vicinity of major hazard sites, such as the petrochemical complex at Canvey Island in Essex.

The term 'material consideration' is similarly broad enough to admit the whole gamut of environmental concerns from visual amenity to pollution. In *Westminster City Council v. Great Portland Estates plc*, Lord Scarman condensed a large body of earlier case law into one general principle which

> applies not only to the grant or refusal of planning permission and to the imposition of conditions but also to the formulation of planning policies and proposals. The test, therefore, of what is a material 'consideration' in the preparation of plans or in the control of development . . . is whether it serves a planning purpose . . . [namely] one which relates to the character of the use of the land.[19]

Despite the generality of the phrase 'character of the use of the land', this judgement (no more than the many others devoted to this question) has certainly not given planners *carte blanche* to devise plan policies, to refuse planning consent or to impose conditions; it simply describes, in terms somewhat less opaque than those used in the text of the statutes from 1947 onwards, the nature and the limits of the discretionary powers conferred upon planning authorities. It implies that, if a planning condition – albeit ostensibly concerned with the social, economic, personal or, more to the point, environmental ramifications of the approved development – can be shown nevertheless to relate fairly and reasonably to the character of the use of the land, then the courts would be unlikely to declare it an abuse of discretionary power.

BOX 3.3 **Development control: consultations**

In order to reduce the risk of planning errors arising through breakdown in communication, the determination of certain planning applications cannot proceed without the views of various state agencies being taken into account. Below is a paraphrased selection of obligatory (a–j) consultations taken from Article 10 of the Town and Country Planning (General Development Procedure) Order (SI 1995 no.419). These are followed by three (k–m) consultations which are 'requested' in the DOE Circular 9/95 which accompanied this order. In both cases, the consultations with a 'hard' environmental remit have been selected.

A local planning authority is obliged to consult the **Environment Agency** in the event of development:

  a involving or including mining operations;
  b involving the carrying out of works or operations in the bed of or on the banks of a river or stream;
  c for the purpose of refining or storing mineral oils and their derivatives;
  d relating to the treatment or disposal of sewage, trade-waste, slurry or sludge;
  e relating to the use of land as a cemetery;
  f involving the use of land for the deposit or refuse or waste;
  g within 250 metres of land which has, within the past thirty years, been used for the deposit of refuse or waste;

the **Health and Safety Executive** in the event of development:

  h within any area notified to the planning authority by the Executive because of the presence of a hazardous (explosive, toxic or inflammable) substance;

English Nature (or the **Countryside Council for Wales**) in the event of development:

  i in or near an area of special scientific interest;

the **Minister of Agriculture, Fisheries and Food** (or the **Secretary of State for Wales**) in the event of development:

  j involving the loss of not less than 20 hectares of grades 1,2 or 3a of agricultural land;

A local planning authority is requested to consult the **Environment Agency** in the event of development:

k   within 500m of a process subject to integrated pollution control;*

the relevant (district council) **Environmental Health Department** in the event of development:

l   within 250m of a process subject to local authority air pollution control;*

the **Health and Safety Executive** in the event of development:

m involving substances notifiable under the Control of Industrial Major Accident Hazards Regulations 1981 (SI 1902) but not requiring a consent under the Planning (Hazardous Substances) Act 1990.

* under Part I of the Environmental Protection Act 1990

Comments from the various agencies which a local planning authority, when determining certain categories of development, is statutorily obliged to consult would undoubtedly constitute 'material considerations'. The categories and the corresponding consultees have been set out in successive General Development Orders; Box 3.3 selects those with the most explicit environmental dimension. Other material considerations of which a local planning authority must[20] take account include policy statements of central government contained in a series of Planning Policy Guidance Notes and in Circulars of the Department of the Environment. These may be very specific in remit or they may outline the role of planning in assisting (or not hindering) the pursuit of the key economic policies of central government. Circular 22/80[21] was issued when the first Thatcher administration's resolve to 'roll back the frontiers of the state' was at its height; it advised local planning authorities that planning consent should be withheld only 'when this serves a clear planning purpose and economic effects have been taken into account'. This presumption in favour of development was couched in the more familiar language of town planning in a subsequent circular:

> There is always a presumption in favour of allowing applications for development, having regard to all material considerations unless the development would cause demonstrable harm to interests of acknowledged importance.[22]

Subsequent reliance, by aggrieved applicants for planning consent, upon this policy of presumption will be assisted by a High Court ruling which gives it the force of law.[23] The case in question happened to be concerned with development in green belt which, despite its historical importance in town and country planning, is what we would label a 'soft' environmental issue.

'Roads versus landscape' issues used to be thought of as 'soft' when compared to emissions from an incinerator or carbonisation plant occupying the 'hard' end of the spectrum. And although different in substance – landscape in Twyford and sulphur dioxide emissions in Monkton – the two case studies which follow both describe attempts by third parties to overcome that presumption in favour of development. What they do have in common is a readiness to exploit opportunities presented by European law in the defence of local environmental objectives.

CASE STUDY 1
Twyford Down

In all important respects, the rights listed in Box 3.1 are replicated in the various quasi-planning regimes which now exist (namely tree conservation, mineral extraction and hazardous substances). The procedures by which major roads are approved have been set down in specific legislation (currently the Highways Act 1980). A number of inquiries into motorways (for instance the M42 at Bromsgrove and the M40 at Warwick) took place before rules of procedure governing inquiries into trunk roads were established. When clarification finally came, it included a rule[24] that the Inspector was to disallow any question 'directed to the merits of government policy'. In one of those cases[25] in which the House of Lords overturned a ruling of Lord Denning in the Court of Appeal, it was later held that the term 'policy' should be interpreted as including the 'need' for a motorway or trunk road. Although jurists came to recognise the need to distinguish between the notions of natural justice applicable to civil litigation between two private individuals and those relevant to administrative decisions directed to the common good, it was the abiding sense of injustice, felt by the early campaigners against the roads lobby, in the dual roles of the Secretary of State for Transport – as both the sponsor and (together with the Secretary of State for the Environment) one of the final arbiters of major road schemes – which ensured that the planning of major roads would never acquire even that grudging sense of legitimacy accorded to the general planning system.

Considerable attention has been paid to the recent 'successes' of UK environmental pressure groups when taking action in public law, but it is necessary to remember that veteran anti-road campaigners have long dismissed third-party objection at public inquiries and similar forms of 'participation' as tantamount to complicity in the violation of the country-side. Fighting for the least objectionable route, saving a copse or a pond, or preserving a Jacobean manor house, they would argue, serves only to deflect arguments from a fundamental debate about transport and perpetuates the Thatcherite belief in an economy based upon the motor car.

To upgrade the existing Winchester bypass section of the A33, rather than have a cutting through Twyford Down, was the aim of Barbara Bryant who, at the 1985 Public Inquiry into an eight-mile (Winchester to Southampton) section of the M3, explained:

> My purpose in objecting is to prevent the undoubted scenic intrusion of the cutting, and to preserve this unique part of our heritage – something to be preserved for future generations whose transport requirements will differ from our own and who will see in a different perspective the desecration of a landscape steeped in the history of [Winchester, the capital of England in Saxon times] undisturbed, and remarkable for its flora and fauna and yet so close to developed centres of population.[26]

The proposed motorway would serve to reduce traffic congestion within the historic city (with its cathedral and college), which accounts for the City Council's initial acceptance of the Department of the Environment's preferred route cutting through the Down. This acceptance was also shared by Twyford Parish Council, the County Council (Hampshire) and the local branch of the Council for the Protection of Rural England. Once evidence presented at the Inquiry left no doubt as to the scale of the full impact of the cutting, and the loss of the sites of archaeological interest,[27] the attitudes of some of these bodies changed. Of more immediate importance was the fact that two statutory bodies had not been represented at the 1985 Inquiry.

Twyford Down formed the western end of the East Hampshire 'area of outstanding natural beauty'; it also included a 'site of special scientific interest' (see Chapter 8); any development threatening such a site required formal consultation with (but not necessarily the approval of) the Countryside Commission. Similarly the two 'scheduled ancient monuments'[28] on the Down necessitated reference to the body now known as English Heritage. The possibility of a judicial review of any decision taken in the absence of these statutory consultations led the Department of Transport to its decision, announced in April 1987, to reopen the Inquiry. In the ensuing four months, Winchester City Council (but not the County Council) reversed its support for the preferred route. Also, the objectors were able to organise themselves into the 'M3 Joint Action Group' and to engage appropriately qualified engineers and landscape architects to prepare the case for their proposed alternative of a tunnel (which they had cursorily suggested at the earlier inquiry).

Evidence presented by senior members of English Heritage's predecessor on the archaeological importance of the site led protestors to speculate on the impact of such evidence, had it been prominent earlier in the planning process.[29] The same was true of the contribution made by the Countryside Commission. Ever since the planning system first acquired statutory duties specific to rural matters, there has been a body – currently the Countryside Commission in England – charged with a responsibility to advise[30] ministers, local authorities and other bodies on matters relating to landscape and natural beauty. At the reopened inquiry, the Commission was represented by a formidable team led by a distinguished member of the Planning Bar. Landscape consultants emphasised the visual impact of the 'canyon' to be cut through the chalk down and contrasted this with the far lower intrusion of their preferred alternative, namely an upgrading of the existing A33. However, this route was vigorously opposed by those (including Winchester College) concerned by the threat it posed to an area of water meadows.

In the event, the advantages of none of the proposed alternatives proved sufficient to dissuade the Inspector from recommending to the Secretary of State that the route (and the cutting through Twyford Down) in the draft order be confirmed.

Thus far the story differs little from many other protests against motorway construction. Those attracted to the more rugged landscape of Dartmoor would probably argue that their failure in 1985 to prevent a bypass to the south of Okehampton, on the fringes of the National Park, represented a greater tragedy. It was the extra-legal action, forming the final coda of the Twyford story, which will be remembered. Civil disobedience is nothing new; but the widespread public sympathy for those protestors who attempted, often at great risk to themselves, to block the path of the earth-moving equipment, was unprecedented. If anything, that sympathy – extending now to social groups most likely to call for stringent penalties for those guilty of 'conventional crime' – has increased with subsequent conflicts, namely the Newbury bypass, the East Devon extension of the A35 and the second runway at Manchester Airport. Media images of phalanxes of police and security staff linking arms to allow bulldozers access have become standard. As protestors have added tunnelling and tree-house construction to their techniques of passive resistance, so the authorities have been required to engage personnel, some with skills in climbing and others in tunnelling, both to serve writs and arrest those subsequently in contempt of court.

The aftermath of Twyford has been so remarkable for environmental action in England that discussion of the protestors' final attempt to pursue their aims by action in the High Court seems almost trivial in comparison. Had they been successful, and forced the government to delay commencement until it had fully honoured its commitments under the Environmental

Impact Directive,[31] then a crescendo of public opinion and political lobbying during this stay of execution might possibly have forced a ministerial change of mind.

Motorways are included in Schedule I of the Directive; and therefore under Article 4.1 an assessment of their environmental effects must be taken into account in the procedures which lead to their consent. Article 5.1 details the information which an assessment should contain; these include 'a non-technical summary'. This information is to be made available to the 'public concerned' (Art. 6.2) and their opinions are to be taken into account in the 'consent' procedure.

Implementation of the Directive in regard to certain public projects, which fell outside the remit of the standard development control system, necessitated specific legislation. For motorways, s.105A was inserted[32] into the Highways Act 1980. One purpose of this addition was to remove schemes, whose draft orders were published before 21 July 1988, from the assessment requirement even though final consent would arise after the deadline for transposition of the Directive.

Twyford Down was one of a number of projects within this 'planning pipeline'. The illegality of the exclusion of pipeline projects formed the first grounds of the Twyford Down Association's challenge[33] to the decision of the Secretaries of State. More specific grounds lay in the failure to supply objectors with a 'non-technical summary'.

Mr Justice McCullogh[34] found himself unable to accept either grounds. On the first, he argued that since the Directive itself made no reference to projects already under consideration at the time of transposition, the United Kingdom government was entitled to assume that they were exempted the Directive's provisions. On the second grounds, counsel for the respondents had argued in court that the report of the Inspector at the 1985 Inquiry could be taken as the 'non-technical summary', notwithstanding its length (322 pages). Without explicitly accepting this contention, he argued that the applicants had failed to refute it. The application therefore fell. However, it is McCullogh, J.'s subsequent *obiter* remarks on standing which have received the most critical attention.

After a lengthy consideration of the ECJ ruling in *Becker*,[35] he accepted that the relevant provisions of the Directive (85/337/EEC) were sufficiently precise and unconditional to have direct effect:

> I have no doubt that the applicants were amongst those whom the directive was intended to benefit and that its provisions were unconditional and sufficiently precise.[36]

However, McCullogh, J.'s interpretation of *Becker* identified an additional requirement before the 'benefit' could be enjoyed:

The first question is . . . to decide what is meant by 'relying upon' rights and 'asserting' rights. I take both expressions to assume that such rights have been infringed. And by infringement I mean that the individual has suffered in some way from the failure to accord him his rights.[37]

It is this notion of having 'suffered' as a precondition of reliance upon the protection implied in the original directive which forms the nub of the issue.

The Court was empowered to quash the decision of the Secretaries of State if the applicants could show that their interests had been 'substantially prejudiced' by a failure to comply with the 1980 Act or regulations made under it. Since the applicants – the Chairmen of the Twyford Parish Council and Compton and Shawford Parish Councils, Barbara Bryant and two other objectors – had not 'suffered' as a result of the faulty implementation of the directive, they could not, Mr Justice McCullogh concluded, enforce it against the state (had their application not been dismissed for other reasons). Exactly what the applicants had not suffered – mental stress, financial loss or physical harm – is not elaborated; but it should be noted that the word 'suffered' does not appear in the relevant part[38] of the 1980 Act.

This *obiter* remark in *Twyford* has been criticised by Alder, who sees it as tantamount to requiring the infringement of a private law right before direct effect can be invoked. There is, Alder argues, 'no reason in principle why public law rights should not qualify for the purpose of direct effect'.[39]

Even though the 'rights' of consultation and participation in the decision-making process conferred by this directive are substantially different from those concerned with taxation in *Becker*, that cannot justify a situation in which national restrictions on remedies are allowed to frustrate their effective enjoyment. Nevertheless, the wisdom of hindsight does feed the suspicion that an action based upon indirect effect – focused less upon the protestors' rights and more upon the disparity between the Directive's aims and the extent of the change in UK practice (as revealed by marginal cases) demanded by the regulations[40] – might have been proven more robust. Had the exhaustion of the protestors' funds[41] not extinguished the possibility of an appeal against McCullogh, J.'s ruling (with the eventual possibility of an Article 177 referral to the European Court), a definitive statement on the direct effect of this important environmental directive might have emerged.

That statement might also have resulted from an Article 169 action against the United Kingdom by the European Commission, acting on the complaint made by the Twyford Down Association in November 1990. The confidentiality that governs discussions between the Commission and a member state during enforcement proceedings is in marked contrast to the transparency which the Environmental Assessment Directive advocates. According to one participant[42] in the Twyford story, the fact that the

Commission could terminate its action against the alleged UK breaches without giving full reasons to the Twyford complainants, casts doubt on the Commission's ability to police effectively any of the aims of the Community's environmental aspirations.

CASE STUDY 2
The Monkton Turbine

Coke had been produced at National Smokeless Fuels' Monkton Works on Tyneside since 1937 when the original bank of thirty-three ovens was built. By 1981 the plant was capable of converting more than half a million tonnes of coal per annum. The carbonisation of coal in the Monkton ovens generated large volumes (approximately 445,000m³ per day) of combustible 'coke oven gas' which, if not immediately utilised as a fuel on or off the site, had to be burnt at a 'flare stack' from which the gaseous products of combustion and sulphur dioxide (arising from the sulphide impurities in the coal) were discharged to the atmosphere.

The Energy Act 1983 entailed certain changes in the rules governing the purchase, by area electricity boards, of electric power from private sources. This erosion in the virtual monopoly of the Central Electricity Generating Board, together with the award of a capital grant by the European Commission under its '1986 Energy Demonstration Projects' scheme, prompted National Smokeless Fuels Ltd (NSF) to consider installing a gas turbine, fuelled by its coke oven gas and capable of generating 3.6MW of electricity, at its Monkton Works (see Figure 3.1). With 70% of this output used on site and the remainder sold at the nationally agreed tariff to the North Eastern Electricity Board, the company stood to save £700,000 per annum.

In August 1986 NSF applied to South Tyneside MBC for outline planning approval for a power generation station (comprising compressor, gas turbine, generator and a 55m chimney) utilising the gaseous products of the carbonisation process. After discussions with the developer, the Town Development Sub-Committee decided on 13 March 1987 to refuse planning consent, for the principal reason that the proposal lacked any

> provision to remove the sulphur content in the process. The proposal would therefore be likely to cause pollution in the form of

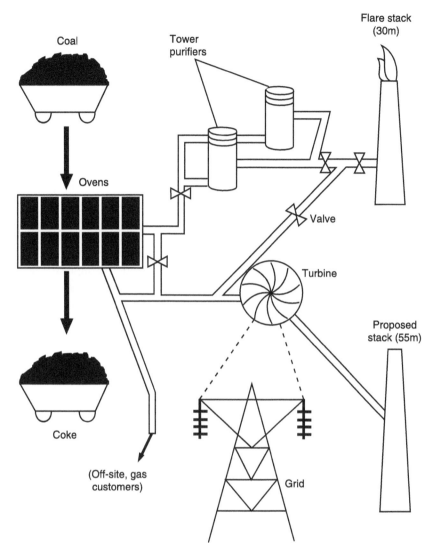

Figure 3.1 Schematic representation of Monkton turbine proposal

acid rain [beyond the immediate vicinity of the works] to the detriment of fauna, flora and human well-being.[43]

In a second proposal, submitted in June 1987, the turbine and its ancillary plant were to be housed in two existing buildings and planning permission was sought for the 55m chimney alone. On 21 August 1987 NSF lodged an appeal against the non-determination of this second application together with an appeal against the refusal of the earlier proposal. The latter, it should be noted, involved no change in the total emission of sulphur dioxide from the works; this would continue to depend simply on the throughput of coal and on its sulphur content. If the turbine were to be operated as planned, two environmental benefits would follow:

1　Some SO_2 which would otherwise have been discharged from the 30m flare stack would now be emitted from a new stack serving the turbine, and being taller (55m), this would afford a small reduction in concentrations in the vicinity of the works.
2　The 3.6MW generated from this source would allow a commensurate reduction in output (and hence in emissions of SO_2 and other pollutants) elsewhere in the CEGB network of coal-fired stations (such as Blyth in Northumberland).

Whilst the local planning authority did not dispute that the proposals would offer a small decrease in local concentrations, they were less impressed by NSF's claim that their scheme entailed a small reduction in overall sulphur emissions. In terms of SO_2 discharged per MW generated, the proposed turbine (4.8gm/sec/MW) compared unfavourably with a typical 2,000MW coal-fired CEGB station (2.6gm/sec/MW) and worse still if such a station were fitted with flue-gas desulphurisation (0.26gm/sec/MW). However, if desulphurisation were to occur *after* the coke ovens and *before* the turbine, then considerable reductions in emissions from the site (and hence significantly lower local concentrations) of SO_2 could be achieved, thereby making discussion of sulphur substitution rates nugatory. A requirement to this effect was included among the planning conditions which South Tyneside urged the Secretary of State to apply if he were eventually minded to uphold the appeal.

At the time of the inquiry the Monkton Works was 'registered' under the Alkali, etc. Works Regulation Act 1906, and Section 5 of the Health and Safety at Work, etc. Act 1974 required the use of best practicable means (BPM) to prevent noxious emissions from both the coke ovens and the proposed turbine. In reaching the decision that BPM for the latter would consist of dispersion of the exhaust gases via a 55m chimney without pretreatment of the turbine fuel (i.e. the coke oven gas), Her Majesty's Inspectorate of Pollution[44] (HMIP) had taken account, the District Inspector explained, of the following:

• the relatively small emission of SO_2
• the relatively small thermal rating (17.5 MW) of the turbine

- the high cost of refurbishing the existing tower purifiers and of alternative means of desulphurising the fuel
- the costs incurred in disposing of toxic wastes arising from any installed desulphurisation plant

This conclusion was reached despite the District Inspector's dissatisfaction with the reliance upon the flare stack and in full awareness of the Inspectorate's published 'Notes on best practicable means'[45] which state that emissions from substantially rebuilt batteries of coke ovens should be 90 per cent desulphurised.

Linking planning consent for the turbine and its chimney with a commitment to remove sulphur from its fuel (i.e. the coke oven gas) became the planning authority's strategy. An awareness that 90 per cent sulphur removal constituted one of the published requirements of BPM for carbonisation works was one of the motivations of this strategy. Another was the authority's knowledge of the existence on the site of plant (the 'tower purifiers') which had been used in the past to remove sulphur (mostly in the form of hydrogen sulphide) from the coke oven gas. The council's aim of securing a significant reduction in sulphur emissions from the existing works was justified by reference to the effects of acid deposition on the wider environment and in terms of various international protocols and commitments, such as that of the so-called '30 per cent Club'.[46] The council's expert witness further argued that although, at 540 tonnes per annum, the sulphur dioxide emitted from Monkton amounted to less than 0.016 per cent of UK emissions, its relative importance would increase as emissions from other sources, especially with the retrofitting of flue gas desulphurisation plant to selected large coal-fired power stations, began to fall. Allowing a new power plant, albeit only 3.6MW (electric), to come on stream without taking the opportunity to reduce its output of sulphur would be viewed with concern by Scandinavian governments.

The solicitor representing the local residents' group indicated that 'coke side arrestment' plant had been promised since the 1981 refurbishment. Its absence was a contributory factor to the appalling record of grit, dust, black smoke, noise and offensive odour which the inhabitants of the nearby housing estate had suffered since 1981, when the current complement of sixty-six ovens came into operation, and which had continued unabated, save for the period of the Miners' Strike in 1984–5, when the cessation of production coincided with a dramatic improvement in the quality of the local environment.

The Inspector was sympathetic to the residents and to their 'unfortunate situation' but the proposal was, he felt, unlikely to worsen their position with regard to dust emission and general nuisance. He also accepted the appellant's claims that the proposal would entail no change in the amount of sulphur dioxide emitted, that discharge via a taller stack could offer a marginal reduction in local SO_2 concentrations and that the proposal entailed the utilisation of a source of energy which would otherwise be wasted. But these considerations were outweighed by those which persuaded him to dismiss the appeal:

> now that the decision has been taken to fit three existing power stations and all new coal-fired power stations with desulphurisation equipment, it is possible to look forward to electricity generation with a reducing output of [SO_2 per MW] of generated power. Although the proposed generating station would not be using coal as its immediate energy source, it is so closely related to coal consumption via the coking process that I think it must be considered in the same category as a coal-fired power station. In my opinion it would be undesirable to permit this new generating station, small as it would be in comparison with CEGB stations, because of its relatively high output of sulphur dioxide per unit of electricity production.[47]

The decision[48] of the Secretary of State to accept his Inspector's recommendations and dismiss both appeals was a little surprising since it ran contrary to the orthodoxy (which was subsequently to be endorsed in the Court of Appeal) that planning controls should not duplicate the powers of regulatory bodies, such as HMIP.

Aggrieved at the unexpected dismissal of the appeal, NSF applied[49] to the High Court to reverse the decision on the grounds that the Secretary of State and, earlier, South Tyneside MBC had acted *ultra vires*. The five detailed grounds cited were concerned, in different ways, with the propriety of determining the planning consent by reference to the level of emissions of sulphur dioxide from the proposed turbine. Of the five, the second:

> There was no evidence before the [Secretary of State] or his Inspector that the decision to install desulphurisation equipment at three existing and all new CEGB coal-fired power stations was a policy which was intended to and did apply to or have necessary

implication for power generating plant of the type and scale proposed.[50]

demands attention, for it was on this point that, in an out-of-court settlement, counsel conceded that the Secretary of State had misinformed himself. Mr Justice Rose duly quashed the decision of 20 June 1988 and awarded costs to the applicant.

Since the point was never debated in open court, we can only speculate on the DOE's response to NSF's fifth and final claim that 'the feasibility and desirability of achieving compliance with a hitherto unannounced policy on emission control' were immaterial and lay outside the scope of the Town and Country Planning Act 1971. However all the issues were to be aired again at the reopened local inquiry.

In July 1988 the gas turbine was installed within existing buildings on the Monkton site, thus planning consent was required only for the proposed 55m chimney. The reopened inquiry into NSF's appeal against the earlier non-determination[51] was held between 30 January and 2 February 1990. In a letter of 13 April 1989, the Secretary of State had indicated to all the parties that he considered the following matters relevant to the redetermination of the appeal:

1   The implications of permitting the emission of unpurified flue gas from the proposed development;
2   the feasibility and potential benefits of requiring the fitting of flue gas desulphurisation equipment;
3   the implications and feasibility of requiring purification of the flue gas before its use to fuel gas turbines;
4   any material change in circumstances which have arisen since the original decision was issued, whether or not it pertains to the matters above.

The view is taken that the only basis on which a condition requiring the fitting of flue gas desulphurisation equipment could be validly attached to a permission for the proposed development would be that without the imposition of such a condition the development would be unacceptable.[52]

Discussion of the first three matters essentially reiterated and refined issues covered at the earlier inquiry. The Residents' Action Group (RAG) argued primarily in terms of the public health implications of works' emissions. Their expert witness eagerly seized upon the recently published study[53] by

Professor Peter Townsend into health and deprivation in the Northern Region. This study, which was a very important contribution to the contemporary discussion of the 'North–South divide' in Britain, revealed that certain wards to the north-east (i.e. downwind of the works) had exceptionally poor health records. The appellants countered this claim by deprecating the uncritical citation of summary statistics which had not been standardised for nutritional, socioeconomic or other factors known to be important determinants of mortality and morbidity. The company's expert witness was Patrick Lawther, Emeritus Professor of Environmental Medicine at the University of London. It was this distinguished epidemiologist whose studies of the health effects of the traditional air pollutants (namely smoke and SO_2) formed the basis of the WHO guidelines[54] as well as the EC Directive on Sulphur Dioxide and Particulates[55] (see Chapter 4).

Given the participation of such authoritative figures, the issues were clearly drawn and the arguments were presented with some cogency. For the most part, this inquiry was in essence no different from countless other 'industry versus environment' inquiries. However new ground was broken when RAG's counsel sought (under the heading 'Any material changes in circumstances') to rely upon the 'direct and indirect effects' of Directive 84/360/EEC.[56]

This Directive is usually referred to as the 'Framework Directive' since it sets out the broad aims (namely major reductions in atmospheric emissions) to be achieved via the terms of authorisations covering emissions from a wide range of industrial sources. It is anticipated that the detailed requirements on, for instance, emission limits for particular pollutants to be observed by a particular industrial sector will be set out in a number of daughter directives. Of the daughters so far conceived, Directive 88/609/EEC on large combustion plant[57] is of passing interest although not immediately relevant since both coke ovens and gas turbines are excluded from its provisions. Annex I of the Framework Directive lists 'coke ovens' among the categories of plant covered; and Annex II cites sulphur dioxide as well as dust and other sulphur compounds among the emissions to be considered in any authorisation.

Counsel for the residents argued that the addition of a turbine and ancillary equipment amounted to a 'substantial alteration' of the Monkton plant which, under Article 3(2), required prior authorisation. Any authorisation which did not entail prevention or reduction of SO_2 (or dust) would ignore the primary purpose of the directive and would be unlawful. Article 4 was central to an understanding of RAG's submission:

> Without prejudice to the requirements laid down by national and Community provisions with a purpose other than that of this

Directive, an authorisation may be issued only when the competent authority is satisfied that:

1   All appropriate preventive measures against air pollution have been taken, including the application of the best available technology, provided that the application of such measures does not entail excessive costs;
2   the use of the plant will not cause significant air pollution particularly from the emission of substances referred to in Annex II;
3   none of the emission limit values applicable will be exceeded;
4   all the air quality limit values applicable will be taken into account.[58]

According to RAG's counsel, the substantive criteria of Article 4 referred to the whole plant, not simply to the alterations which would trigger such an authorisation. The Inspector was further advised that the wording dictates that the competent authority must withhold an authorisation until the applicant had demonstrated that 'all appropriate preventive measures' had been taken and that 'significant air pollution' would not be caused. The existence on the site of the tower purifiers and the recognised availability of other desulphurisation plant meant that NSF would be hard pressed to claim that 'all . . . measures' had been taken. If they invoked the 'not entailing excessive costs' caveat, it should be understood that, if costs were adjudged to be excessive, this would be due to NSF's improvidence in allowing the water cooling towers to fall into disuse.

Counsel for the residents outlined the EC case law from the three principles of supremacy, sympathetic interpretation and direct effect stemming from Article 5 of the Treaty. From the Inspector's record of the cases cited, it is apparent that a combination of the last two was central.

Formal compliance with the Directive was required by 30 June 1987. That date had passed and, because a substantial alteration to an industrial consent could be authorised with 'no mention of prevention or reduction of pollution', it was apparent that the directive had not been fully implemented into national law. Therefore the residents were entitled to rely upon direct effect in a UK court or tribunal (including a planning inquiry) to secure the benefits which the directive was intended to confer upon them.

Irrespective of the non-implementation of the Directive, the doctrine of sympathetic interpretation was relevant in this case. The Secretary of State (and by implication the Planning Inspector who advised him) was an emanation of the state and was obliged to consider any relevant Community law when exercising any quasi-judicial role (such as the determination of a planning appeal) under domestic law. Fulfilment of the general aims of Directive 84/360/EEC dictated, if not dismissal of the appeal, that substantial pollution reduction measures, namely desulphurisation of the coke oven gas, should be incorporated as planning conditions. This obligation was fundamental; it existed irrespective of the opinions of HMIP and of the absence of specific daughter directives.

The recommendations of the Inspector, which were endorsed by his Assessor, can be stated very briefly. He recommended that consent for the proposed chimney be given subject to the standard conditions on timing plus a condition requiring the fitting of suitable equipment to desulphurise the coke oven gas. The relevance of RAG's lengthy submission on the application of EC environmental legislation was referred to the Secretary of State, whose decision can be stated with similar brevity: he allowed the appeal and granted planning consent for the 55m chimney, subject only to the condition that building commence within five years. As for the Community law principles:

> neither Article 5 nor Article 130r of the Treaty of Rome confers rights on individuals which can be enforced in national law.[59]

His discretion in determining this appeal was not circumscribed because:

> Directive 84/360/EEC applies to coke ovens but, as the Commission of the [EEC] has been formally notified, its provisions are presently implemented by the Alkali, etc. Works Regulation Act 1906, the Health and Safety at Work, etc. Act 1974 and Regulations made under these Acts.[60]

The Resident's Action Group, and later South Tyneside MBC, gave notice of their intention to challenge the legality of this (second) decision of the Secretary of State. However the cessation of operations at Monkton and the demolition of the plant meant that the High Court was not required to give what might have been the first ruling on the 'direct effect' of an environmental directive.

If, as the decision letter of 28 March 1991 contended, the directive had been adequately implemented by the 1906 and 1974 Acts, then it must follow that it became somewhat more than adequately implemented after 1 January 1991,[61] when Part I of the Environmental Protection Act 1990 came into force and 'best practicable means' finally gave way to 'best available techniques not entailing excessive cost' (BATNEEC) as the driving force of UK air pollution control. It is otiose to speculate on the attitude of the High Court in the abortive appeal. Irrespective of the particular domestic legislation which was held, at the time of the minister's decision, to implement Article 4(1), it is difficult to see how 'dispersion via a tall stack' could be taken as an adequate response under the new regime of integrated pollution control. However, an argument based upon sympathetic interpretation does seem more compelling than direct effect. There can be little doubt that the framework directive translates the policy aim – to secure a substantial reduction in emissions to atmosphere from industrial sources – into Community law. Article 1 clearly states the purpose of the Directive to be

> to provide for further measures and procedures designed to prevent or reduce air pollution from industrial plants.[62]

In determining planning appeals, a minister acts in a quasi-judicial role. In the period before those 'further measures' are enacted, he must use the discretion which that appellate role entails so as to assist the 'prevention or reduction of pollution' and, in this case, to anticipate the strict emission limit which the relevant daughter directive would inevitably require. To suggest otherwise would seem *prima facie* to be contrary to the principle of sympathetic interpretation set out in *Von Colson*[63] and subsequently in *Marleasing*:

> in applying national law, whether the provisions concerned pre-date or post-date the directive, the national court asked to interpret national law is bound to do so in every way possible in the light of the text and the aim of the directive to achieve the results envisaged by it.[64]

Of course, had the residents brought this case to court, their statutory review might have fared no better than that of the Twford Down Association. This case does at least offer a model for other environmental groups to follow, once they can identify a breach of relevant Community law, which is not crucially dependent upon a recognition of rights.

Conflicts

The problem of overlapping jurisdictions is not new. A draft planning circular[65] in 1972 laid stress on the need for early consultation with HM Alkali and Clean Air Inspectorate[66] during the planning process. The draft circular particularly deprecated the practice of applying planning conditions to 'scheduled works' which then became more stringent controls than those demanded by the (central government) Inspectorate. Publication of the circular was suspended when the Royal Commission on Environmental Pollution announced its intention to review the whole question of air pollution legislation. The 1976 Report of the Commission[67] generally endorsed the pragmatic tradition of the Inspectorate and was scathing in its opposition to duplicative planning conditions. However, this 'problem' was soon to be overshadowed by the far greater one of reconciling the UK approach to air pollution with the very different requirements (especially the reliance upon quantitative standards of air quality) of the European Community (see Chapter 4). Planning Policy Guidance no. 23 therefore represents the latest in a series of attempts to reduce the conflict between planning and pollution control authorities (HMIP at that time):

> Lack of confidence in the effectiveness of controls imposed under pollution control legislation . . . is not a legitimate ground for the refusal of planning permission or for the imposition of conditions on a planning permission that merely duplicates such controls.[68]

The determination of planning consent for sources of pollution has tended to be complicated by two issues: differences between lay and scientific perceptions of the risks posed to human health, and duplication of regulation. Both were central in the Court of Appeal's[69] ruling over the Gateshead incinerator.

A proposal to build a clinical waste incinerator was opposed by the local planning authority for reasons which included a concern over the impact on air quality and public apprehension of the effects of dioxin emissions. After an inquiry, the Planning Inspector held that these were sufficient grounds for refusing consent and recommended that the applicant's (Northumbrian Water Group) appeal be dismissed. The Secretary of State did not accept that recommendation. In justifying his decision to grant planning consent, he relied principally upon the existence of the statutory controls over emissions from the incinerator which would be exercised by HM Inspectorate of Pollution. The legality of this justification was challenged by the local planning authority. At the High Court hearing, the deputy judge presiding held that whilst the threat of harm to the environment and public health was undeniably a 'material consideration', so too was the existence of a control regime statutorily obliged to address those threats, and it was for the Secretary of State to consider their relative weights and to decide accordingly. This view was subsequently endorsed by the Court of Appeal, where Glidewell, L. J. went on to make an observation which could have implications not merely for incinerators but for a wider range of environmental problems:

> Public concern is, of course, and must be recognised by the Secretary of State to be, a material consideration for him to take into account. But if in the end that public concern is not justified, it cannot be conclusive. If it

were, no industrial development – indeed very little development of any kind – would ever be permitted.[70]

This ruling serves as a reminder, if any were needed after Twyford Down, that the planning system is driven by calculations of aggregate utility, with the desires and preferences of third parties, no matter how firmly held, counting for little if they cannot be translated into tangible reasons for negating the presumption in favour of development.

4

DO WE HAVE A RIGHT 'TO BREATHE CLEAN AIR'?

Introduction

One of the advertisements on behalf of Friends of the Earth which has appeared in the UK press asserts that 'clean air is a basic human right'. Clean air was cited first in the list of 'legally enforceable environmental rights' which the UK Labour Party's Policy Commission on the Environment committed a future Labour administration to recognise; and disputes over air pollution would fall within the jurisdiction of an 'Environment Division' of the High Court, the creation of which formed another commitment in this policy document.[1] In this chapter I consider the extent to which rights-based arguments have featured in British and US approaches to the control of air pollution. These approaches are many and various, with few being entirely devoid of a rights dimension. They range from action in nuisance as a means by which a householder can secure remedies against smoke and fumes originating from neighbouring land, to international treaties on global pollution. Given the exigencies of atmospheric dispersion, the political objective of 'clean air' requires intervention at the local, regional and supra-national levels, with varying opportunities for involvement by individuals and citizens' groups.

In Hohfeld's terminology (see Chapter 1) a right to breathe clean air is the correlative of the duty not to pollute the atmosphere. But if those enjoying the rights are indistinguishable from the bearers of the duties, this is little more than a tautology. A more fruitful distinction may be found between mechanisms which attempt directly to protect the (substantive) right to breathe clean air and those which enable individuals to participate in various forms of intervention to reduce polluting emissions to the atmosphere. Again, a clear distinction cannot always be drawn; but any 'right to clean air' must embrace the latter even though popular understanding tends to concentrate upon more direct remedies such as tort.

Among those venerable cases in tort referred to in Chapter 1 are those which established that smoke, grit and dust from industrial sources could give rise to nuisance whether in terms of injury to property[2] or on grounds of interference with domestic tranquility and personal comfort.[3] Such interference must be 'unreasonable', with this term understood to take account of the nature of the locality: 'what would be a nuisance in Belgrave Square would not necessarily be so in Bermondsey'.[4] In the original judgement in *Rylands v. Fletcher*, Blackburn, J. indicated that 'fumes and noisome vapours' could give rise to action under the rule which he was then articulating.[5]

Private actions against sources of air pollution still occur today, but they are far fewer than actions in statutory nuisance taken by local authorities. Indeed the enforcement of public health legislation of the last century constituted one of the very *raisons d'être* of local government in the UK. It is worth noting however that with the most recent amendment of the law of statutory nuisance, the private individual still retains a power[6] to bring a complaint before the magistrates. The interference with the enjoyment of domestic life in mid-nineteenth century England was, in essence, no different from that suffered by Mrs Lopez Ostra in modern Spain (see Chapter 1), where the failure of the state to strike a reasonable balance between the conflicting interests of the individual and the community has been adjudged to amount to a violation of a human right.

In the first section of this chapter, attention turns to those fixed sources of industrial air pollution which, by virtue of technical difficulties in their control, have fallen within the jurisdiction of a specialist agency of central government, currently the Environment Agency. Given its importance, we must pay some attention to Part I of the Environmental Protection Act 1990, even though it is not a particularly rich source of procedural rights. We then consider the way in which European law might give rise to something approaching a substantive right to breathe clean air. Finally, instances of a residual rights perspective in managerial and economic approaches (where they might be least expected) to air pollution are identified.

The regulation of industrial air pollution

Until 1991, the application of the 'best practicable means' (BPM) to prevent discharges to atmosphere from 'scheduled processes' had formed the cornerstone of the statutory regulation of industrial air pollution in the UK since 1874. Private prosecutions for failure to use 'best practicable means'[7] were possible in theory, but the consent of the Attorney General, or later the Director of Public Prosecutions was required. The author is not aware of any such actions nor of any private prosecutions under Part I of the Environmental Protection Act 1990, where this restriction does not apply. The history[8] of the enforcement of the Alkali Acts is not one of zealous application of statutory controls; rather it is one of confidential negotiation between regulator and industrialist, with only occasional involvement of the legislature (when new processes were added to the 'schedule') and with the role of the citizen confined to that of a complainant (usually in vain) against the noxious vapours from various works using one or more 'scheduled processes'.

The 1906 Act made one concession to the role of private individuals: s.22 provided that if any sanitary [i.e. local] authority, on information given by any of their officers, or any ten inhabitants of their district complained to the central authority (eventually the Secretary of State for the Environment) about nuisance occasioned by a works 'scheduled' under the 1906 Act, then the central authority was obliged to 'make such inquiry as . . . they think fit and just'.

Records of such inquiries do exist; perhaps the last of the species was that held on 27 July 1971 into the complaint by Northfleet Urban District Council concerning the persistent failure of Associated Portland Cement Manufacturers Ltd to operate its electrostatic precipitators so as to reduce dust emissions from its Northfleet plant.

> Intermittent operation was the greatest bugbear and the company was diligent in its endeavours to settle down to steady operating conditions.[9]

Nevertheless, the Inspector, Mr R. G. Adams, an Assistant Secretary at the newly formed Department of the Environment, concluded that 'best practicable means had been used' and the residents were offered only sympathy:

> It was fully appreciated how much local residents, especially those bordering the works, had suffered during the erection, commissioning and early teething troubles of the plant.[10]

Liaison committees, involving residents, local councillors and industrialists, have been set up in response to persistent problems with a number of works. But the absence of opportunities for more formal involvement in the enforcement of the technological controls over some of the sources of greatest nuisance must, at least in part, account for the readiness of some local authorities to use powers conferred upon them as planning authorities to this end (see Chapter 3 for an example).

Under the current successor to the Alkali Act 1906 (i.e. Part I of the Environmental Protection Act 1990) there is no equivalent to the s.22 inquiry. Local authorities are no longer confined to the role of initiator of an inquiry; in fact they and the Environment Agency now enforce a common[11] regulatory regime over fixed sources of air pollution. Regulations distinguish the more technically demanding processes, to be regulated by the central body, from those which fall within the jurisdiction of local authorities; but the principles – especially the use of the 'best available techniques not entailing excessive cost' (BATNEEC) – underlying Part I of the 1990 Act apply to both. Like BPM, BATNEEC entails a residual duty on the operators of certain processes to prevent the release of substances into the environment and, when prevention is not practicable, to reduce any release to a minimum and to render it harmless. This duty represents an implied condition on all authorisations (under s.6 of the 1990 Act); other specific conditions may be imposed in pursuit of various objectives including compliance with European Community obligations[12] or with international treaties, the achievement of quality standards or of any national 'plan'.[13] Failure to observe BATNEEC and the violation of a specific condition can both lead to enforcement action. Should a future dispute generate case law, it would be equally binding on both the central and local jurisdictions.

When taking action, whether in the civil or criminal courts, the right of access to the public register (s.20 of the 1990 Act) and the information on any prescribed process would be of considerable value to a private individual. It follows that an exclusion of information, on the grounds of commercial confidentiality, would considerably impede such action. Access to information on pollution discharges has seen a remarkable change in official attitudes over the last two decades. The principle of a public register – giving information on each consent to discharge (liquid effluent into surface waters), any conditions attached, and more importantly the results of any monitoring which could reveal any breaches of that consent – was established in Part II of the Control of Pollution Act 1974 (COPA). But implementation of this legislation was very dilatory; and it was the Royal Commission on Environmental Pollution, and especially following investigations described in its tenth report,[14] which became the most influential advocate for change. The 'public interest' grounds for exemption from the register should be confined to 'national security'. In addition, the Royal Commission called for an extension of COPA principles to be extended to all pollution control regimes with the exclusion on grounds of 'commercial confidentiality' used only in the most exceptional cases.[15] In regard to industrial air pollution, the Royal Commission recognised that this would require an amendment of the statutory regime[16] which at that time forbad disclosure of any data on discharges to atmosphere.

More generally the Royal Commission advocated a shift towards a system, comparable with that in the USA, in which all information was available to the public as of right, unless the regulator or polluter established convincing reasons why it should be withheld. Had the Royal Commission not adopted such an unequivocal stance, it is possible that compliance with the 1990 directive on access to information on the environment[17] might have proved more difficult. In the event, there are still some industrialists who defend the need for commercial confidentiality exemptions by arguing that publication of information on the chemical composition of the discharges from their chimney (or liquid effluent pipe) could enable a rival company to gain an insight into processes and substances which otherwise remain 'trade secrets'.

Table 4.1 refers to numbers of appeals that were generated by the initial authorisations (under Part I of the 1990 Act) and given the obligation[18] to review all authorisations at least once in every four years, opportunities would appear to arise for individuals and environmental groups to involve themselves in the regulation process. The latest regulations[19] allow any public representations, made in regard to an authorisation or an appeal, to be included in the public register, but private individuals are not entitled to take part in the appeal, to question any written submission or to cross-examine appellants or regulators at a hearing. There is of course the option of judicial review for those residents and others who, following the recent relaxation in judicial attitudes, can satisfy the 'sufficient interest' criterion.

Table 4.1 Air pollution appeals, 1 April 1991–31 March 1997

Enforcing agency:	*Local authorities*		*HMIP*	
	s.15[a]	s.22(5)[b]	s.15[a]	s.22(5)[b]
Lodged	67	6	72	6
Withdrawn	34	2[c]	61	2
Outstanding	11	0	7	0
Upheld	17[d]	3[e]	3	3[f]
Dismissed	5	1	1	1[g]

Source: Private communication, Air and Environment Division, Department of the
Environment.

Notes:

[a] 1990 Act, against refusal of an authorisation or against conditions.

[b] 1990 Act, against refusal to withhold information from public register on grounds of commercial
confidentiality.

[c] These two were in fact 'out of time'.

[d] Of which six were upheld in part.

[e] Of which one was upheld in part.

[f] Shell, National Power, ICI.

[g] PowerGen.

An attempt by a resident of Pembroke to challenge the authorisation of
National Power's power station in that Welsh borough has been reported.[20] At one
stage HMIP itself had advocated the use of 'integrated gasification and combined
cycle technology' (IGCC) as BATNEEC for this plant, where 'orimulsion'[21] is
burnt. Therefore a challenge of the eventual decision to accept flue gas desulphurisation or other techniques to minimise emissions could hardly be dismissed
as a trivial one by a vexatious litigant. After viewing HMIP's affidavit the applicant
chose not to pursue a formal challenge. His request for the opportunity publicly to
debate the extent of HMIP's duty to give reasons for its decisions was rejected by
the Court.

In the event, the first public law actions associated with the 1990 Act were taken
by local authorities. In two cases, a local authority challenged the Secretary of
State's decision to reject his Inspector's recommendation and uphold the appeal,
by the operator of an animal rendering works, against refusal of authorisation.[22]

There is no recorded instance of a refusal (or of a consequent appeal) of regis-
tration under the Alkali Acts. Thus the controversy surrounding the authorisation
of the Point of Ayr Gas Terminal is evidence of the change in the attitudes of the
central body. Situated in North Wales, this terminal forms part of a £1 billion
investment to exploit Britain's west-coast oil and gas reserves. A planning inquiry
was held in 1992, at which HMIP, in the absence of finalised proposals for pollution minimisation, declared itself unaware of grounds to object to the proposal.

The Secretary of State for Wales granted planning consent in February 1993. But when HMIP finally received an s.6 application in May 1994, it was unable to agree that the operator's proposals for limiting atmospheric emissions of sulphur dioxide and other pollutants[23] amounted to BATNEEC. This particular dispute was eventually resolved without recourse to legal action.

The manufacture of cement is a 'prescribed process' and regulated by the Environment Agency. The very high temperatures (1,400°C) necessary for the chemical changes which produce the cement can exceed those found in hazardous waste incinerators in which toxic wastes, including chlorinated hydrocarbons,[24] are destroyed. The use of industrial waste as 'secondary liquid fuel' burnt in cement kilns represents one of the most contentious issues to have arisen under the new regime of the 1990 Act. It has been conducted in a very open manner, attracting the critical attention of the House of Commons Environment Committee;[25] but the environmental and legal implications are so many and varied that resolution of this latest dispute could well involve action in the High Court.

Distilling and re-using methanol and other substances is widely accepted as the environmentally optimal option for treating contaminated industrial solvents. Burning the residues from this process in cement kilns reduces the amount which goes to landfill – the least acceptable disposal option in terms of sustainability. Proponents of the use of solvent residues as an auxiliary kiln fuel argue that it entails less pollution than 100 per cent reliance upon coal or petroleum coke: a claim which seems justified in terms of reduced emissions of oxides of sulphur and nitrogen, but needs to be examined in more detail in regard to a number of heavy metals.[26] Merchant incinerator companies who (for a fee, typically in the region of £90 per tonne) collect these compounds and burn them at high temperatures in purpose-built plant with no energy recovery are very much concerned at the threat posed to their business by what, for the waste producers, is a cheaper option (typically £30 per tonne).[27] They too need to utilise the calorific value of solvent residues in destroying toxic material in sludges and in aqueous solutions. They stress that these substances remain 'waste' even though their calorific value can reduce the amount of coal or other primary fuel consumed in cement manufacture. The precise legal designation of 'secondary liquid fuel' (SLF) has a number of consequences.

The transport of this 'waste' (as with all other forms of controlled waste) would be subject to the 'duty of care'.[28] More importantly, the kiln could come within the remit of the EC Directive on Hazardous Waste Incineration.[29] If the waste provided more than 40 per cent of the total calorific value, it would be necessary for the kiln operators to meet the stringent emission control standards applied to waste incinerators under that directive. However, the important question is whether the use of such fuel changes the planning status of the kiln (to a waste incinerator) and thereby requires planning consent. If planning permission is required, there arises the possibility of a planning inquiry (following an appeal over refusal or conditions or a ministerial 'call-in' of the application) whereupon local residents and environmental groups have the opportunity of participation. Comparable opportunities are not offered by the authorisation process under Part I of the 1990 Act. Hence the planning status of the use of SLF is of considerable importance to residents (and members of planning authorities) who would have objected vehemently to planning consent for a cement kiln had they known that it might subsequently be used to incinerate industrial wastes.

Burning SLF offers the opportunity for substantial savings which could and should, according to HMIP's interpretation of BATNEEC, be used to fund improved particulate arrestment – echoing an earlier decision to allow the burning of 'orimulsion' at National Power's 2,000MW oil-fired power station at Pembroke[30] only on condition that the savings were used to retrofit flue gas desulphurisation plant estimated to cost £300–400 million. As for the concerns of the merchant incinerator companies, the Chief Inspector has commented:

> [HMIP] is not required to keep the incinerators in business or to make sure that any sector records a profit. . . . It's not our role to keep any sector in business to the detriment of the environment.[31]

As a measure of the change in attitudes of the central inspectorate, it is instructive to compare this comment of the Chief Inspector with one made by a predecessor thirty years earlier:

> The country's, industry's and work's current financial situations have to be weighed against the benefits [of reduced emissions to atmosphere] for which we strive and careful thought has to be given to decisions which could seriously impair competitiveness in national and international markets.[32]

Francovich liability and EC directives on air quality

The air quality limit values for smoke and sulphur dioxide set out in Directive 80/779/EEC[33] were implemented in Great Britain by regulations[34] which required that the limit values on smoke and sulphur dioxide finally be met by 1 April 1993. Violations of the mandatory limit values are now rare; so much so that it might be argued that the official data might be more informative if, instead of the limit values, they took as a benchmark the more stringent 'guide values' which member states are under a general duty to move towards. Data produced by the Department of the Environment[35] revealed that violations of the limit values in the United Kingdom (outside Northern Ireland, where special circumstances apply) tended to occur in towns in or near coalfields, where the benefits to miners of concessionary (bituminous) coal dampened the local authorities' enthusiasm for the introduction of smoke control areas.[36] However, with the closure of so many mines in recent years, far fewer exceedences would now be expected to arise.

In the great majority of cases of the violations occurring in the 1980s, it was the limit value for the 98 percentile (in effect, the value which appears seventh in the year's daily averages listed in descending order of concentration) for smoke which was exceeded. The limit values for smoke and sulphur dioxide are not independent. Figure 4.1 presents the 98 percentiles calculated from data recorded at a monitoring site at Hetton-le-Hole (a mining village in the north-east of England and one of the British districts subject to a derogation from the original 1983 compliance date of the directive) and, for comparison, data from an urban site in nearby South Shields. The use of the 98 percentile is a recognition of the existence of 'episodes' in which meteorological conditions can lead to short periods of very high pollution. The near violation of the SO_2 limit in South Shields ($249\mu g$ m^{-3} in January 1982) is probably an example; but it was still a pale imitation of the notorious London smog of 4–8 December 1952, when low temperatures and anticyclonic conditions conspired to generate far higher concentrations, leading to an excess mortality estimated at around 6,000 in south-east England.[37]

Such conditions are now eradicated in Great Britain. Belfast (in Northern Ireland) is one of very few remaining areas where a time series plot of the maximum hourly concentration of sulphur dioxide (see Fig 4.2) still shows a distinct annual periodicity, with the peaks coinciding with midwinter anticyclonic conditions offering poor dispersion of emissions from the coal burnt, for domestic heating, in open hearths. The far less regular pattern shown in the plot from the monitoring site in Central London is evidence of the greater role of transport sources. Despite the overall decline in concentrations in recent decades, respiratory distress can still be caused by the high levels which can arise over short time periods; for that reason the World Health Organisation has set standards for one-hour- and ten-minute averages. A link between respiratory morbidity and daily changes in the levels of air pollution was explored in a study[38] conducted in the vicinity of the Monkton coke works (to which reference is made in Chapter 3). Here the epidemiologists were able to identify a statistically significant trend between daily SO_2 concentrations and the proportion of respiratory-related consultations recorded in general practitioners' surgeries in the immediate vicinity of the coking works. This study was unable to identify a similarly significant trend with smoke concentration.

The 'six cities' study in the United States was far more impressive in its scale, aims and methods.[39] It examined mortality rates in six American cities displaying the range in pollution characteristics which exists within that continent. A prospective cohort study, it ended in 1991 and involved over 8,000 white subjects over the age of twenty-five when the study began in 1994. Information was recorded on a variety of potentially confounding factors (age, weight, educational attainment, medical and occupational history, smoking habits) which were then used in adjusting the mortality ratios. The results are summarised in graphical form in Figure 4.3, from which the strongest association (when the points corresponding to each city lie on an almost straight line) is shown by the graph referring to fine particulates, namely those with an aerodynamic diameter of less than

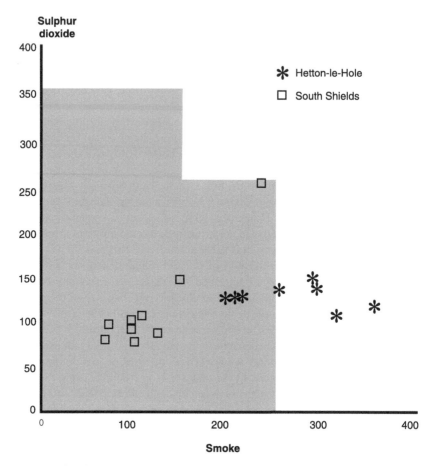

Figure 4.1 Smoke and sulphur dioxide: 98th percentiles at two sites in NE England
(1981–90)

Note:
Using data from Warren Spring Laboratory's network of smoke and sulphur dioxide monitors, the 98 percentile of the (365) values of 24-hour average smoke (horizontal axis) and sulphur dioxide (vertical axis) concentrations for each year from 1981–90 are plotted on a graph. By drawing the 98 percentile limit values given in EEC/80/779 (*viz.* LV$_{smoke}$ = 250; LV$_{SO_2}$ = 350, if 98 percentile smoke <150, or LV$_{SO2}$ = 250, if 98 percentile smoke >150; the units are µg m^{-3}) as lines on this graph, exceedences are then identified by points lying outside the shaded area. The exceedences recorded at Hetton-le-Hole (a mining village in NE England) appear as the four points lying to the right of the graph in the area indicating excessive smoke. For comparison, equivalent data from a monitoring site in South Shields, an urban centre some thirty miles to the north, are also plotted.

2.5 microns[40] which enables to them escape the lungs' various defence mechanisms and to penetrate into the lining of the air sacs where the greatest physiological damage is caused. The graph for sulphate (constituents of acidic deposition formed by the oxidation of sulphur dioxide in a variety of chemical

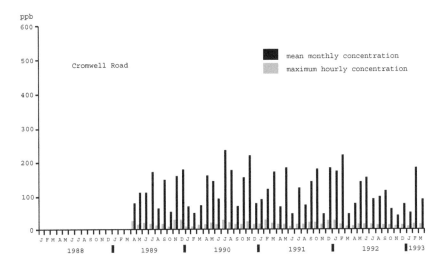

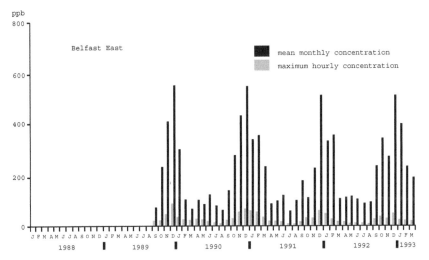

Figure 4.2 Sulphur dioxide concentrations at two contrasting (Belfast and London) UK
sites: January 1988–March 1993

Source: Quality of Urban Air Review Group (1993) First Report: Urban Air Quality in the
United Kingdom (QUARG, London).

reactions in the atmosphere) is similarly significant. The authors argue that their
findings are consistent with other studies which have found associations between
air pollution and rates of hospitalisation, decrease in lung function, asthma attacks
and other respiratory symptoms.

There is no evidence pointing to 'threshold' concentrations for particulates
(below which their contribution to excess mortality, chronic and acute morbidity

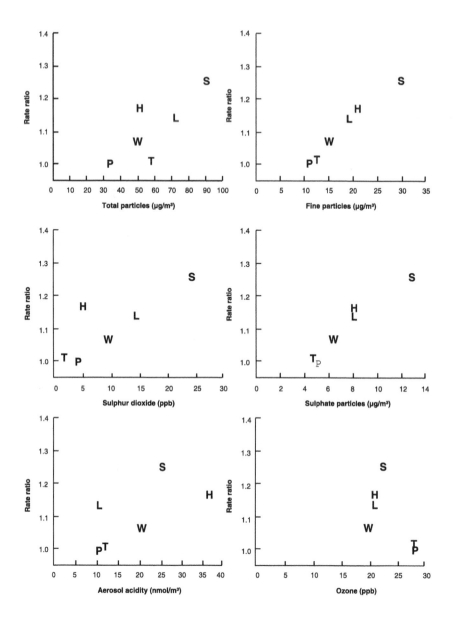

Figure 4.3 Estimated adjusted mortality-rate ratios and pollution levels in six cities in the USA

Source: D. W. Dockery et al., 'An association between air pollution and mortality in six US cities', New England Journal of Medicine, 329 (24) (1993) 1754.

Note: P = Portage WI   L = St Louis MO
       T = Topeka KS   H = Harriman TN
       W = Watertown MA   S = Steubenville OH

ceases). Since it is not possible to identify a concentration at which the state's duty to protect respiratory health is discharged, efforts to reduce urban air pollution can always be justified. But whenever a regulatory regime is based upon quantitative environmental quality objectives then – in a manner totally alien to the British tradition – the breach, or fulfilment, of the state's obligation to its citizens may be indicated by the location of points on a graph. Therefore, the choice of limit values becomes crucial. For example: the 'guide value' (i.e. a maximum 24-hour concentration of particulates to be within the range $80-120\mu g$ m^{-3}) cited in 80/779/EEC was exceeded on at least one day in each of the ten years between 1980–9 at Hetton-le-Hole, and on at least one day in seven of those years at South Shields. Had the epidemiological opinion then demanded the adoption of this lower figure as the mandatory standard, then the UK government would have been faced with non-compliance as the norm in far more areas.

These issues would be of more than academic interest were failure adequately to implement a directive like 80/779/EEC ever to be the subject of *Francovich*[41] litigation in an English court. Exceedence of the smoke limit value might plausibly be said to arise out of the Secretary of State's failure to use his discretionary power[42] to direct the local authority to initiate a smoke control programme. If, as Ludwig Krämer and others (see Chapter 2) have argued, Directive 80/779/EC is one which, having the protection of health as its subordinate aim, confers a right upon individuals, then anyone who can demonstrate harm as a result of such an exceedence would appear to have grounds for seeking compensation from the state.

It is difficult to imagine an environmental case in which the facts might bear parallels with those that led to the original *Francovich* decision. Once Signor Francovich had demonstrated his standing, it followed (indeed it was almost tautologous) that he had suffered a financial loss in circumstances which the original directive was designed to mitigate. Does exceedence of an air quality directive's limit value *ipso facto* imply harm, or does the burden of proof remain with the plaintiff? The limit (and guide) values were motivated by a concern with human health and the actual values were based upon recommendations of the World Health Organisation; but the third condition of *Francovich* liability (i.e. a causal link between the breach of the state's obligation and the damage suffered) would seem to favour the latter view. In which case, do the limitations of epidemiological knowledge of respiratory morbidity confine standing to those with pre-existing conditions such as bronchitis, emphysema or asthma? Who exactly can claim to have suffered from the exceedence indicated by data from monitoring station X? If I live closer to station Y, whose data have never exceeded the limit values, is my standing thereby negated? These questions might require resolution by the European Court if such an action in *Francovich* were to be contested by the member state.

When the state breaches its duty to protect the atmosphere – by ensuring compliance with a particular limit value – the classical problem of identifying the offender from a multiplicity of pollution sources no longer arises. But it is replaced by the problem of a potential multiplicity of pollution victims, each seeking compensation for respiratory morbidity. In this regard, Directive 85/203/EEC is another fruitful object of speculation. It sets a limit value (98 percentile of hourly mean values, recorded throughout the year, of $200\mu g$ m^{-3}) for nitrogen dioxide. Petrol-engine vehicles form the principal source of this respiratory irritant, which is also the raw material in the photochemical production of ozone (see below) and many other toxic pollutants. Official statistics[43] would suggest that no exceedence of the limit value for nitrogen dioxide has been recorded in the 1990s; however, the equivalent guide value (98 percentile of hourly mean values, recorded throughout the year, of 71ppb or $135\mu g$ m^{-3}) is still exceeded in parts of Central London. Another nitrogen dioxide guideline (the WHO recommendation of a four-hourly concentration £50ppb) was exceeded on 142 days in 1994.[44]

The possibility of those millions (or even that minority who are prone to asthma) who live or work in Central London seeking *Francovich* damages from HM Government in respect of harm to health from exposure to nitrogen dioxide concentrations in excess of the EC limit value is a consequence of the view that this directive (or an amended version) confers a right upon individuals. Whether a directive of this type confers a right which can be enjoyed in practice depends primarily upon the level at which the (mandatory) limit value is set.

BOX 4.1 Stopping the traffic

Section 14 of the Road Traffic Regulation Act 1984 empowers a traffic authority to restrict or temporarily ban vehicles from using any road within its jurisdiction for reasons which include 'the likelihood of danger to the public'. A fire or similar emergency might necessitate closure of a road at short notice irrespective of the inconvenience to its users. An action group of residents near Trafalgar Road (A206) in south-east London sought to persuade the traffic authority (the London Borough of Greenwich) to use this power whenever temperature inversions raised concentrations of pollutants from vehicle exhausts to the point of exacerbating asthma suffered by children living in the vicinity. In 1995 a group of parents, acting on behalf of their children, applied to the High Court for a declaration that air pollution could amount to a 'danger to the public' justifying temporary road closure.[45] Mr Justice MacPherson had no hesitation in dismissing the application, arguing that the provision in question had to be construed in the context of 'road traffic matters' which, contrary to the claims of the applicants' counsel, could not be stretched so as to embrace occasional episodes of high pollution.

It is difficult to take issue with this ruling. In its aims and its unsuccessful outcome, it is similar to *Budden* (see Chapter 1) and it reinforces the lesson of that earlier case, that the health risks from vehicle emissions can only be reduced by controls at source (i.e. fuel content, compulsory fitting of catalysts and prosecutions for excessive smoke). However, the advent of a system of 'road pricing', if given sufficient flexibility to increase tolls at short notice, might offer effective incentives to use public transport during the worst anticyclonic conditions. But again, the problem of motorists attempting to avoid the extra costs by diverting to minor roads (not covered by the pricing system) has to be faced.

If the prospect of mass claims seems nonsensical – would claims from asthmatic motorists be dismissed on grounds of complicity? – it does at least remind us that, in urban centres at least, pollution with the most adverse effect on public health is predominantly emitted from mobile rather than static sources. It follows therefore that a 'right to breathe clean air' cannot easily coexist with motorists' interpretation of the right to liberty – the right to drive wherever and whenever they wish (see Box 4.1). Recent moves by the UK government towards a more 'managerial' approach to air quality have been accompanied by a greater recognition of the contribution made by transport sources.

The management of air quality

The latest directive involving the health effects of an atmospheric pollutant is concerned with ozone. This (tri-atomic) form of oxygen is often referred to as a 'secondary' pollutant since, at ground level, it is produced in photochemical reactions – involving principally nitrogen dioxide and carbon monoxide (another component of car exhaust gases) – and forms an important ingredient in the summer smogs which afflict most cities. The ozone directive[46] does not lay down mandatory limit values as in those air quality directives promulgated in the 1980s. It defines two reference concentrations: one defining a level (an hourly average of 90ppb) above which the public are to be 'informed'; and another (an hourly average of 180ppb) above which they are to be 'warned' of the health dangers posed by exposure to such levels of this pollutant.[47]

The Council of Ministers agreed the text of a European Community directive on air quality assessment and management in June 1995. This directive will serve as a framework for twelve daughters, each dedicated to a particular pollutant, of which the existing directive for ozone is the first. The innovation of information and warning thresholds will be extended, but the mandatory limit values for others will be retained. However, a consultation paper published in 1994 suggested that the UK government would be unlikely to support any strengthening or extension of the current regime of mandatory standards on ambient concentrations of atmospheric pollutants. In a wide-ranging review of UK air quality, the Department of the Environment's support for standards was less than wholehearted:

in spite of valuable work by the European Community and [World Health Organisation], there are grounds for doubt whether existing standards and guidelines will provide a coherent basis for understanding the effects of pollution, developing action to combat it, and guiding air quality management in Britain and Europe as we move into the 21st century.[48]

In January 1995 a document[49] was published outlining the Department of the Environment's plans for:

- a framework of revised air quality standards and targets
- local air quality management areas
- effective control of emissions from vehicles

The formulation of revised air quality standards and targets has been the responsibility of an 'Expert Panel on Air Quality Standards' (EPAQS) set up by the Secretary of State for the Environment in fulfilment of a commitment included in *This Common Inheritance*. EPAQS is charged with recommending standards for nine atmospheric pollutants of most significance in terms of their effects on human health and the environment. These standards will form part of a National Strategy which the Environment Agency will take into account when setting the conditions[50] of authorisations of prescribed processes. Local authorities will carry out periodic reviews to identify areas where air quality targets are unlikely to be met. Plans will then be drawn up to improve air quality in the most cost-effective manner. Further information on the measures local authorities may take to secure these improvements are contained in consultation papers; one is concerned with local authority involvement in traffic management.[51] As well as measures to encourage walking, cycling and the use of public transport, local authorities can assist in traffic management so as to reduce congestion and encourage the freely flowing conditions in which emissions are lowest. In extreme cases, certain categories of vehicle (e.g. heavy lorries) might be banned from specified routes in the interest of air quality.

In all these documents published in the aftermath of the Environment Act 1995, the emphasis is not upon the right to breathe clean air, but upon the cost-effective management of air quality. In any involvement by the Environment Agency, its officers are under a statutory duty 'to take into account costs which are likely to be incurred and the benefits that are likely to accrue'[52] by that involvement. And while the protection of health is recognised as the primary motivation of that managerial role by the Agency and other organs of the state, there is no allusion to individual rights or grounds for compensation for health detriment attributable to some failure of that role.

Economic approaches

When passing the 1970 amendments to the Clean Air Act, the United States Congress declared that every citizen has a statutory right to be protected from 'any known or anticipated adverse effects',[53] but the subordination of this lofty ideal to the exigencies of economic realities has been well described by Mackay.[54] He argues that the US Environmental Protection Agency, although a product of the 'rights revolution' of the 1960s, has been forced to adopt an approach much closer to that combination of pragmatism and technological feasibility which characterised the pursuit of 'best practicable means' in the UK. This change was due in part to the willingness of the executive to accept the arguments of industrialists on the crippling costs of the abatement technology necessary to meet the EPA's air quality standards. Any 'command and control' regime – of which the UK's enforcement of BPM was an example – depends upon the regulatory authority having information on the abatement costs. In the real world, such information is never completely available and one of the advantages of the alternative 'economic incentives' (see Box 4.2) approach is that its absence is not crucial.

BOX 4.2 Pollution permits

Suppose that a firm is permitted to discharge sulphur dioxide (or any other pollutant) up to a specified maximum (expressed in tonnes per annum). Suppose also that, if through improved clean-up technology, it discharges less than its quota, then the residue could be sold to a firm whose production targets cannot be met without exceeding its quota. A market in quotas will develop: firms with low pollution per unit of output can increase their profits either by selling their surplus or by increasing production after purchasing extra quotas; others would be forced to 'clean up' or go out of business. The 'polluter pays' principle is satisfied; the cost of pollution is internalised along with other costs (raw materials, labour, finance) and a form of economic efficiency, which does not treat the atmosphere as a free waste disposal resource, becomes achievable.

But for proponents of the 'minimal state', the main attraction of a system of marketable permits or quotas is that it can operate with the regulatory authority having a smaller role. That body must calculate the total annual emissions, having regard to the relevant atmospheric quality limit values. It must also decide the initial allocation of those emissions. If it were prepared to accept the political implications of immediate bankruptcies, it might do this by offering discrete units for sale at a grand auction. If industry opposition obliges it to set quotas in line with existing emissions, it is 'in effect capitalizing the implicit right to pollute'[55] which those sources enjoyed under the preceding regime. After the initial period, market forces take over and little intervention by the regulatory authority is required. By buying and selling quotas, it could mimic the role of a central bank; by accumulating its own stock of quotas, it could assist a progressive improvement in overall atmospheric quality towards some 'guide value'.

The traditional reliance upon tall stacks – offering sufficient dispersion of untreated emissions to ensure that national (or state) air quality standards for oxides of sulphur and nitrogen are not exceeded – can increase acid deposition further afield. A desire to tackle the source of the problem contributed to the 1990 decision of the US Congress to amend the Clean Air Act so as to enable market-oriented approaches to achieve a national target of sulphur dioxide emission ten million tons below the 1980 level. Under s.404 of the Act, the Environmental Protection Agency must allocate 'marketable pollution allowances' to fossil-fuelled power plants; this initial allocation was on the basis of each plant's past emissions and fuel consumption rate. In what was reported to be the first transaction in the newly created pollution permit market, the 'clean' Wisconsin Power and Light Company sold the right to emit 10,000 tons of sulphur dioxide, at a price of $300 per ton, to the 'dirty' Tennessee Valley Authority, some 1,000 miles distant. The purchaser claimed it was simply spending a small sum buying a little time before its own long-term $0.75 billion investment would lead to emission reductions of the order of 800,000 tons by the end of the decade.[56]

The UK's conversion to economic instruments of pollution control, as revealed in *Our Common Inheritance*,[57] is relatively recent. The increased excise duty on leaded petrol and the tax per tonne of waste sent to landfill (see Chapter 7) are examples of monetary disincentives to abstain from environmentally damaging activities. The UK opposed an EC proposal for a tax based upon the carbon content of fuel.[58] But it is possible to point to a UK emission control system which could lead to marketable quotas.

Under the Large Combustion Plant directive,[59] the UK undertook to reduce its total emissions of sulphur dioxide and nitrogen dioxide. From the reductions actually achieved by 1994 (see Box 4.3) it would seem that the targets are not unduly demanding. This has not been lost on other northern member states of the European Community, who recall the UK's resistance to this directive and the fact that the UK's reductions were eased to reflect its exceptional reliance upon coal-fired generation.[60] Environmentalists have been further angered by the fact that the reduction has achieved more by the introduction of North Sea gas-fired generators rather than the promised amount of flue gas desulphurisation of existing coal-fired plant.

The LCP requirements resulted in the import from the USA of the 'bubble' concept. The Directive applies to combustion plants with a power rating in excess of 50MW; such sources fall naturally into three categories, namely electricity generation, refineries and other industries (e.g. metal works). Each of the (now privatised) generating companies is set an overall limit (or bubble) for its total emissions of sulphur and nitrogen oxides; the refineries and other industry are also given a bubble. Trade-offs between the different plants within each bubble are permitted provided that no unacceptable impact on local air quality results. There was a report of National Power agreeing 'to transfer'[61] 12,000 tonnes of its 1995 surplus SO_2 allocation and 17,000 tonnes in 1996 to Northern Ireland Electricity to assist the Ballylumford B station in the period preceding its conversion from coal

BOX 4.3 Proposed reductions in UK emissions of sulphur dioxide and nitrogen oxides

Unless otherwise indicated, the data below indicate the percentage reductions on the 1980 emissions from 'Large Combustion Plant' (LCP) as required under Directive 88/609/EEC.

	1980	1993	(1994)	1998[a]	2003
	ktonnes	%	%	%	%
SO$_2$	3,883	-20	(-49)	-40	-60
NO$_X$	1,016	-15	(-45)	-30	

Note:
[a] the percentage (1980=100%) reductions in emissions actually achieved in 1994

Following the signing, in Oslo on 14 June 1994, of a protocol to the 1979 Geneva Convention on Long-range Transboundary Air Pollution[*] (under the auspices of the UN Economic Commission for Europe) the requirements (again expressed as percentage reductions on the 1980 emission) for UK all sources of sulphur dioxide emission in the first decade of the next century were set as follows:

	1980	1990	2000	2005	2010
	ktonnes	ktonnes	%	%	%
SO$_2$	4,898	3,780	-50	-70	-80

More recently, the European Commission's proposal to use 1990 emissions as the baseline for fixing even more drastic reductions in both SO$_2$ and NO$_X$, as laid down in the EC's fifth Environmental Action Programme, has been criticised by the UK and other member states who claim that they would thereby be penalised for their prompt compliance with earlier demands for curbs on acid gas emissions. The tough limits on NO$_X$ emissions would undermine the rationale of Britain's latest generation of power stations fuelled by (low sulphur) North Sea gas, which were built partly in response to earlier obligations to reduce national emissions of SO$_2$.

[*] International Legal Materials (1985) 24, 484.

to gas. There is no record yet of the actual sale of all or part of a quota. Having denationalised electricity generation and introduced the 'bubbles', the trading of quotas would seem the next logical step; but two obstacles have appeared.

First, each power plant must still secure authorisation[62] under the Environmental Protection Act 1990, BATNEEC must be observed and the Environment Agency will not allow local air quality to be unduly compromised or any 'critical load' to be exceeded. The critical load of a particular pollutant (e.g. acid deposition) for a given receptor (e.g soil) is defined as the highest deposited load which that receptor can withstand without incurring long-term damage.[63] Consideration of critical loads introduces a locality dimension into the quota discussion. Although an extra one hundred units emission from one plant (and the corresponding reduction at another) might be consistent with economic efficiency, it might be disallowed if the impact of that increased emission upon an adjacent area caused a critical load to be exceeded. The proximity of a number of sensitive Welsh 'sites of special scientific interest' (SSSIs; see Chapter 8) was reported to have influenced the decision to demand the retrofitting of flue gas desulphurisation at the Pembroke power station burning orimulsion.[64] Given the geographical distribution of the SSSIs in the UK, power plants situated in Northern Ireland, Scotland and north-west England pose a threat, in terms of the acid deposition on the soil of these sites, which is disproportionate to the total sulphur emissions[65] from these plants. If protection of SSSIs takes priority over a power generator's 'right' to emit an otherwise agreed sulphur quota, then ecocentric considerations (see Chapter 8) can longer be dismissed as fanciful and irrelevant to the practicalities of the regulation of major sources of air pollution.

Second, there is still some unfinished business in regard to privatisation: the Office of Electricity Regulation is reported[66] to have threatened a referral to the Monopolies and Mergers Commission unless the two major generators agreed to dispose of part of their inherited capacity. Existing emissions quotas would have to accompany any disposal in order to allow the purchaser the freedom to increase generation. The potential for oligopoly in this market is clearly apparent from Table 4.2. Reconciling these other requirements whilst retaining a 'free market' in pollution quotas could prove an insuperable problem for the UK regulators.

In the USA the word 'right' is often used as a synonym for permit or quota. But in reality, the use of quotas to assist an economically efficient level of ambient air pollution and the enjoyment of a right to breathe clean air are not easily reconciled. It could be argued that a system of marketable quotas in the UK would serve to perpetuate that right to pollute (or that immunity from prosecution) which, from 1874 until 1991, was enjoyed by those whose discharges defied arrestment by the best practicable means. Does the adoption of a marketable quota system leave any role for the individual or environmental group?

In theory, an emitter could buy so many permits and exhaust them so rapidly that local residents suffered nuisance. But even if the regulatory agency did not intervene, the threat of civil action would be a further incentive to efficiency, and there seems little grounds for arguing that immunity from such action should be

Table 4.2 Large combustion plants: pollution emissions and allocations in 1994

	SO_2		NO_x	
	E	A	E	A
Power stations:				
National Power	893	1,373	261	401
PowerGen	730	969	176	247
Scotland	64	102	41	59
Northern Ireland	71	75	20	20
Sub-total:				
Power	1,758	2,519	498	727
Refineries	75	99	27	32
Other Industry	136	267	48	114
Total	1,969	2,885	573	873

Source: *ENDS Report* (1995) 248, 11.

Notes:
E = emissions (μ *tonnes*)
A = allocations (μ *tonnes*)

conferred upon polluters who participate lawfully in a permit scheme. It is possible to imagine circumstances in which a residents' group might challenge the legality of an initial allocation to a given source. A group campaigning for clean air might wish to challenge the price per tonne or the overall tonnage of, for example, nitrogen dioxide to be bid for at the inaugural auction. But once the scheme is operating as the economists and legislators envisaged, it is difficult to identify a role for private individuals.

Conclusions

Urban air pollution is now characterised by multiplicities of both polluters and pollutees. As such, it falls within that category of environmental problems which would appear to lend themselves to the 'superfund' approach (see Chapter 7 below). Whilst such schemes have tended to be concerned mainly with the long-term chemical contamination of land, there is no reason in principle why they cannot be used to compensate victims of chronic[67] respiratory morbidity. In fact, a Japanese fund,[68] although better known for its compensation of the victims of 'Minamata disease',[69] can be used by residents, in areas prone to high ambient concentrations, to pay for medical and related expenses. The onus of proof, and the reliance upon epidemiological evidence, does not arise since the fund compensates only for the effects of pollution which cannot be attributed to a specific polluter. 80 per cent of the fund comes from the operators of stationary sources of sulphur dioxide and 20 per cent from a tax on vehicles. Whether the legal right of individuals meeting the qualifications to receive compensation falls within the ambit of a wider interpretation of 'environmental right', rather than the more familiar 'welfare right', is open to question. To apply the same label to the right of UK citizens to medical treatment (under the National Health Service) for respiratory disorders, which may or may not be pollution-related, seems an even greater semantic distortion.

When considering this notion of a 'right to breathe clean air', the distinction which Dworkin has drawn between rights and goals seems particularly relevant. A right, he argues, stems from a political principle, whereas

> a goal is a non-individuated political aim, that is, a state of affairs whose specification does not . . . call for any particular opportunity or resource or liberty for particular individuals.[70]

Clean air, like other welfare aspirations, is best understood as a goal. Given that the atmosphere is a common property resource, history suggests however that the achievement of that goal entails the extinction, rather than the extension, of individual rights.

All European directives impose obligations on member states; from some, like 80/779/EEC on smoke and sulphur dioxide, it is possible to infer 'rights' conferred upon individuals. An individual has a legitimate expectation that an obligation to protect the atmosphere and hence public health should, like any other, be honoured. But an expectation does not necessarily entail a right. It is hard to escape the conclusion that UK subjects will need something more than a provision in a European directive before they could truly enjoy a right to breathe air free of that cocktail of pollutants which urban dwellers currently have to inhale.

5

A RIGHT TO CLEAN WATER?

Introduction

Water is an essential requirement of life itself. Any attempt to codify fundamental human rights which lacked a statement of a right to water (and food) would be seriously flawed; any state which purports to respect human rights has a duty to ensure that its citizens have access to water, sufficient in both quality and quantity, to meet their physiological needs. In a sub-Saharan African state, that duty will consist primarily of ensuring access to vitally adequate resources, especially in periods of drought. But in all but the most inhospitably arid regions this undeniably 'environmental' resource is one which falls from the sky, and therefore prolonged denial of water rarely occurs. In the context of a modern industrial state situated in temperate latitudes, that duty must be discussed in terms of the economic factors which determine the quality of water supplies. Since chemical and biological contamination of water can compromise its use for any purpose, not simply that of human consumption, consideration of a state's duty cannot omit reference to cleanliness. Among the other purposes, the recreational opportunities afforded by surface and coastal waters are perhaps the most sensitive to pollution.

If the impossibility of 'owning' the atmosphere gives it the characteristics of a 'common property resource', it is tempting to assume that the oceans, rivers and lakes are similarly definitive examples.[1] But oceans can be owned in the sense that any maritime state possessing, and willing to use, a sufficiently powerful navy can control access to, trade within, and the exploitation of marine resources in, the seas within the geographical limits of its influence. That state may choose to forego the sovereignty which its naval supremacy affords; and it may freely enter with others into agreements by which navigation, fisheries and mineral extraction are regulated by some supra-national authority.[2]

There may be a sense in which maritime states have chosen to regard the seas, outside their territorial waters, as 'commons', but at the other end of the hydrological cycle, private ownership remains firmly entrenched. The rights and privileges that attend the ownership of even the smallest watercourses have been recognised and protected for centuries, not by naval power, but by the common law. A riparian owner's entitlement to 'water of his stream in its natural flow, without sensible diminution or increase, and without sensible alteration in its character or quality'[3] has a claim to be recognised as an 'environmental right' even if the term was not in common or legal parlance in 1893. Since a riparian right is best understood as an extension of traditional 'property rights', it does not amount to a *quintessential* environmental right (see Chapter 1).

The pollution of a stream, the use rights of which are clearly and unambiguously assigned, can become the subject of bargaining between polluter and pollutee, enabling an economically efficient arrangement (in which the combined cost of pollution damage and pollution arrestment is a minimum). Ronald Coase's seminal contribution[4] was to recognise that this optimal outcome was independent of any particular initial allocation of legal rights and duties (see Figure 5.1 and Box 5.1). The simple two-party dispute, perhaps most closely realised in the reported cases by *Young & Co. v. Bankier Distillery Co.*, rarely arises today. In practice, a multiplicity of downstream victims of pollution, each with a different incentive to bargain, causes Coasean bargaining to break down. In addition, the costs incurred by the negotiations themselves mean that one particular allocation of entitlements becomes more conducive to economic efficiency than another. As Calabresi has argued,

> a decision will often have to be made on whether market transactions or collective fiat is more likely to bring us closer to the Pareto optimal result the 'perfect' market would reach.[5]

'Collective fiat' seems far too harsh a phrase to describe the loose regime of statutory controls which, whilst leaving civil remedies intact,[6] gradually brought various acts or omissions which are now labelled 'water pollution' within the scope of criminal law in the United Kingdom. With the Industrial Revolution, abstraction from rivers and their use as a free waste disposal service expanded to such an extent that the state was obliged, for reasons more concerned with the protection of public health than the pursuit of economic efficiency, to acquire the power to permit or refuse discharges to river.

The Rivers Pollution Prevention Act 1876 created an offence where any person caused to fall or flow or knowingly permitted to fall or flow or to be carried into any stream any poisonous, noxious or polluting liquid proceeding from any factory or manufacturing process. The use of 'best practicable means' (as with the earlier controls over industrial smoke) was a defence; and proceedings by private individuals could only be taken with the consent of the Local Government Board. This body (the central government department exercising certain functions over local authorities) was directed not to give its consent to proceedings which might inflict 'material injury [on] manufacturing industry'.[7] The river authorities, appointed to enforce the Rivers (Prevention of Pollution) Acts of 1951 and 1961, were no longer subject to such restrictions; but any private individual wishing to prosecute under these Acts had to secure the consent of the Attorney General.

BOX 5.1 Coasean bargaining and riparian rights

Imagine a discharger of pollution, a paper mill for example, situated upstream of another commercial activity, such as a trout farm, which depends upon clean water. The polluter (paper mill owner) is aware of the cost he incurs in preventing each successive unit of pollution from the maximum Z down to 0; in other words, he can calculate the marginal control cost curve (MCC). The pollutee is similarly aware of the cost (profits lost through increased fish mortality) he incurs as the result of the damage caused by each successive unit of pollution from 0 to the maximum Z; in other words, he can calculate the marginal damage cost curve (MDC).

Riparian rights assigned to trout farm owner

In this arrangement, the interest of the party holding riparian rights lies in demanding clean water. This is achieved by the mill owner reducing his effluent to zero, thereby incurring a control cost 0ZG. He can reduce his total cost by persuading the trout farmer to accept a tax per unit pollution discharged. If the tax is set at 0T per unit discharged, he then incurs a total cost comprised of 0TAP in tax and APZ in reducing pollution from Z to P (not 0); this is still less (by an amount GTA) than he would be obliged to pay in the absence of bargaining. The trout farmer now suffers some (0AP) damage but, with the tax revenue, he is still better off (by amount 0TA); in other words, he has an incentive to enter the bargain. By inspection of the higher Y and lower V amounts, it is clear that pollution level P minimises the mill owner's costs whilst maximising the trout farmer's net incentive to bargain.

Riparian rights assigned to paper mill owner

Now riparian rights are enjoyed by the party whose interest lies in using the river as a free waste disposal service. He will discharge Z units of pollution thereby incurring 0 control costs. The trout farmer therefore has a damage cost 0ZH, but he can reduce this by persuading the mill owner to accept a subsidy for each unit of pollution NOT discharged. If, again, the subsidy is set at 0T per unit of pollution foregone, then the trout farmer pays a subsidy of APZF and still incurs a damage cost of 0PA; but this total is still (AFH) less than he would incur in the absence of bargaining. The mill owner must pay APZ in control costs but this is less than the subsidy he receives, leaving a net incentive (AFZ) to enter bargaining. Again, by inspection of the higher Y and lower V levels, it is apparent that P minimises the trout farmer's costs whilst maximising the mill owner's net incentive to bargain.

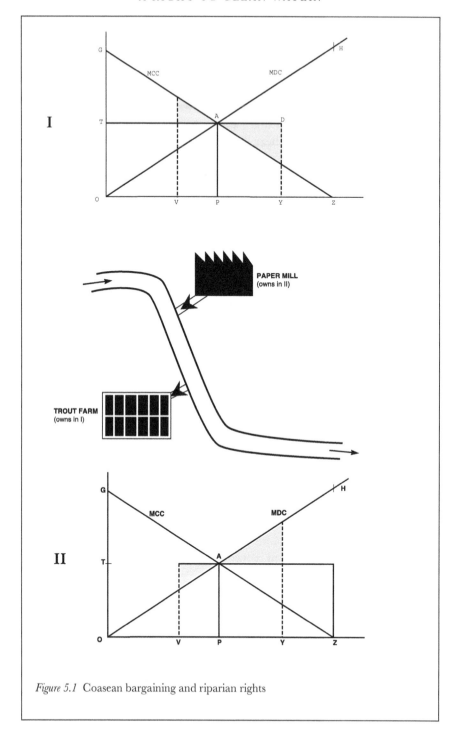

Figure 5.1 Coasean bargaining and riparian rights

Now ignoring both taxes and subsidies and considering the social costs (i.e. the sum of the total damage and control costs), it can be shown that position P (where MDC and MCC intersect) defines a minimum, namely economic efficiency. As Coase first realised, the achievement of economic efficiency is independent of the original assignment of riparian rights: this simply dictates whether the bargaining involves a tax or a subsidy to reach this position. Provided that the bargaining process itself entails no costs, and if each party pursues self-interest fully aware of his own (but not necessarily the other's) marginal cost curve, then a mutually acceptable result will ensue.

In practice, there are reasons why Coasean bargaining is of limited application:

1 Bargaining (litigation) can be costly;
2 there will be a number of downstream pollutees, each with a different incentive (i.e. a different marginal damage curve) to bargain;

in addition:

3 'good planning' would situate polluters, not as shown in the diagram, but downstream of pollution-sensitive users thereby removing the need for bargaining or, for that matter, regulation.

The choice of 0T was not incidental: setting the tax or subsidy level above or below 0T does not obviate bargaining, but it does lead to departures from economic efficiency.

This formalism can be used to illustrate the incentive to developing a more cost-effective control technology (the MCC inclined more to the horizontal) to be operated by the mill owner. Both tax and control cost saving follow from a lower cost per unit of pollution controlled. In fact inefficiency can arise unless the tax level is reduced to take account of the new technology. Parallel arguments apply if the farm owner were to introduce more pollution-tolerant strains of trout.

Further reading

Burrows, P. (1979) *The Economic Theory of Pollution Control*, London: Martin Robertson.

Coase, R. (1960) 'The Problem of Social Cost', *The Journal of Law and Economics*, 3, 1–44.

Sandbach, F. (1982) *Principles of Pollution Control*, London: Longman.

In 1973, the twenty-nine rivers authorities, 157 water undertakings and 1,393 sewerage authorities were succeeded by ten multi-purpose Regional Water Authorities (RWA) in England and Wales. The chairmen and the majority of the members of these bodies were appointed by the Secretary of State; the remainder were elected members of the various local authorities within the area served by the particular water authority. One may question the extent to which RWAs were democratically accountable, but they were clearly public bodies, over which central government (in effect the Department of the Environment) exercised ultimate control (via appeals over discharge consents, appointments, loan sanctions, etc.).

But it was the economic policy rather than the executive power of central government which prevented RWAs from pursuing a more rigorous enforcement against industrial polluters. For most of the brief lifetime (1973–89) of these authorities, stringent curbs on public sector expenditure were imposed by both Labour and Conservative administrations. They had inherited, from local author- ities, crumbling sewers and such inadequate waste water treatment plant that the water authorities themselves were major polluters of rivers and coastal waters within their areas. Prosecution of others, in all but the most flagrant circum- stances, would lay them open to the charge of double standards. Part II of the Control of Pollution Act 1974 required a public register to be kept, listing the conditions imposed on all consents to discharges to 'controlled' waters. This provi- sion was accompanied by the removal of the remaining statutory restrictions on private prosecutions. Proceedings were occasionally instituted by environmental groups and by angling societies, but their impact on overall standards of water quality was minimal.

Privatisation of the water authorities necessitated the vesting of their regulatory powers within new bodies, the National Rivers Authority (NRA) and the Drinking Water Inspectorate (DWI) which, like Her Majesty's Inspectorate of Pollution, were specialist units within the Department of the Environment. The NRA, before its assimilation into the Environment Agency, proved very willing to prosecute water companies whose discharges to rivers had violated the conditions of their particular consents. Although it was the NRA's prosecution[8] of Shell UK (for contaminating the River Mersey with 30,000 gallons of crude oil) and the resulting fine of £1 million which received most publicity, it was this greater readiness to prosecute statutory undertakers which represented the most climactic change.

The period that witnessed the rise and fall of the regional water authorities also saw the growing influence of the environmental programmes of the European Community. Despite early conflict between the United Kingdom's flexible and pragmatic tradition and the Community's concern with the quantitative stan- dards, the European environmental programme is the primary determinant of regulation of the important sectors of the aquatic environment in all member states. The five directives discussed in detail below have been selected (from a large number concerned with various aspects of the aquatic environment) because of their particular relevance to a rights discourse. In the UK, compliance with

European Community directives has necessitated massive investment to make up for decades of neglect of infrastructure, from sewers to water treatment plant. Whether or not it is meaningful to speak of a right to drink or to swim in clean water, the effective enjoyment of either right does not come cheaply.

Five European Community directives relating to water

Groundwater

The movement of water underground, and the dispersion of any pollutants within that water, is indifferent to any pattern of ownership of the land above. Aquifers perhaps have more of the characteristics of a 'common property resource' than surface streams. Their long-term value as a source of potable supplies is so much greater than, and so easily destroyed by, their use as a sink for liquid wastes that stringent criminal penalties to deter such misuse is now accepted. Since the contamination may only be apparent years after the offence, the difficulty of identifying the offender tends to inhibit prosecution. The regulatory role of the state is primarily one of prevention. For member states of the European Community, where aquifers form the principal source of drinking water, that role is now largely circumscribed by a comprehensive set of directives. As with the equivalent directives protecting the atmosphere, those concerned with the aquatic environment offer similar scope for discussion, if not actual instances, of direct effect and *Francovich* (see Chapter 2).

The 'groundwater' Directive[9] seeks to protect these valuable natural resources from pollution by toxic chemicals (e.g. mercury, cadmium, certain pesticides and many organohalogen compounds). In an Article 169 action over Germany's defective implementation of this Directive, the European Court explicitly stated that the Directive creates rights for individuals:

[80/68/EEC] seeks to protect the Community's groundwater in an effective manner by laying down specific and detailed provisions requiring the member states to adopt a series of prohibitions, authorization schemes and monitoring procedures in order to prevent or limit discharges of certain substances. The purpose of those provisions of the Directive is thus to create rights and obligations for individuals.[10]

That these rights may be conferred upon environmental groups as upon private individuals is apparent from part of the opinion of the Advocate General:

Clear and precise implementation of the Directive's provisions may also be important for third parties (for instance environmental groups or neighbourhood residents) seeking to have prohibitions and restrictions contained in the Directive enforced as against the authorities or other individuals.[11]

An annex to this directive specifies (in List I) substances, the *direct* discharge of which into groundwater must be prevented. This prohibition is precise and unconditional and, as the Court affirmed, it confers rights. According to Krämer[12] it is therefore capable of direct effect. In contrast, member states must take 'all appropriate measures they deem necessary to prevent any *indirect* [i.e. after percolation through ground or soil] discharge'[13] of these substances; the discretion implied in this requirement means that indirect discharge does not entail direct effect. Similarly, List II contains those less dangerous substances, whose introduction into groundwater is to be, not 'prevented', but 'limited' – a term implying a sufficient measure of discretion to deny direct effect.

A direct discharge of a List I substance into an aquifer would therefore constitute a breach of the Directive. Since this Directive confers rights upon individuals, and the content of this right is determinable from the text of the Directive, any individual who can establish a causal link between damage he has suffered and a direct discharge would appear to have grounds for a *Francovich* action against the state. However, it is difficult to imagine circumstances in which a plaintiff would be forced to rely solely on this largely untested right of action. If the individual had a right to abstract water from an aquifer polluted by any substance (irrespective of any breach of the Directive and whether it appeared on List I or List II) civil action under various heads, including the rule in *Rylands v. Fletcher*,[14] could be pursued. If the individual possessed no particular property right with regard to the aquifer, his right to drink clean water from the aquifer might be said to have been infringed in a manner giving rise to a *Francovich* action. In such circumstances, a directive concerned specifically with drinking water (see below) would seem a more promising source of remedy. In fact, *Cambridge Water Company v. Eastern Counties Leathers*,[15] one of the most important English civil actions in recent years and very much concerned with the contamination of an aquifer, made only passing reference to European law. But it did lead to the important ruling by the House of Lords that the 'foreseeability' of damage should now be regarded as a prerequisite of liability under the rule in *Rylands v. Fletcher*. The Cambridge case is discussed at more length in Chapter 7 below.

Cadmium discharged into the aquatic environment

In the days before the precautionary principle forbad us to think of the oceans as sinks with an almost infinite capacity for our wastes, UK regulatory bodies sought to take advantage of Britain's position as an island in the North Atlantic swept by the Gulf Stream and cleaned daily by its many tidal estuaries. Putting their trust in dilution, they had an antipathy to any control regime based upon uniform standards imposed on each point of emission irrespective of the quality of the receiving medium.

The Framework Directive[16] for regulating toxic discharges to the aquatic environment contained a compromise whereby each member state could choose between one of two regimes. For each of the more toxic substances (specified in List

I, which is not quite identical to that included in the Groundwater Directive), a limit value would be specified in a daughter directive; and the standard to be met by each source of emission was to be at least as stringent as this specified value (for example, a discharge from the manufacture of cadmium compounds must not exceed 0.5 grams of cadmium per kilogram). In the alternative regime (favoured by the UK alone) the permitted emission for each source would be calculated by reference to a quality objective also specified in the daughter directive (for example, the total cadmium concentration in any inland surface water must not exceed $5\mu g$ per litre). This framework directive was one of the first to be promulgated following UK accession to the Community. It is difficult to imagine a similar concession to the UK being offered today. However, the condition of the environment in some of the former Eastern Bloc countries seeking membership may necessitate many transitional arrangements, if not a distinct shift towards subsidiarity.

The Groundwater Directive (above) is one in which, according to the European Court of Justice,[17] the intention of conferring a right upon individuals is readily apparent; it is also a daughter of the Framework Directive. It is ironic therefore that one of its sisters – the cadmium directive[18] – should be the subject of a later ruling of the ECJ which casts doubt upon the direct effect of many other environmental directives. But the *Luciano Arcaro*[19] case is far from straightforward. Italy, despite its failure to implement the Cadmium Directive, was attempting to use it to prosecute a discharger who was not obviously in default of the existing domestic law. This is tantamount to an inversion of direct effect, and the Court had no compunction in dismissing Italy's (Article 177) request for clarification. However, the opinion of Advocate General Elmer raises issues of wider significance. Without entering a discussion of rights of individuals, he appears to suggest that the Cadmium Directive, by allowing a competent authority to set discharge limit more strict than that specified in the text of that Directive, entails sufficient discretion to cause it to fail the 'unconditional' test for direct effect.[20] Whether this ruling represents a fatal blow to the direct effect of those many other environmental directives, which allow member states to impose limits in excess of some Community-wide minimum, remains to be seen. If the possibility of a member state exercising that choice is sufficient (by making a directive less than 'unconditional') to deny citizens the opportunity to remind their member state of its obligations, it most certainly undermines Krämer's 'estoppel' interpretation (see Chapter 2) of direct effect. It further encourages those who have suffered harm as a result of faulty implementation of the cadmium or any cognate directive to seek redress via *Francovich* rather than direct effect.

Bathing water

The 'bathing water directive'[21] sets nineteen physical, chemical and microbiological standards to be met in areas of sea and freshwater where bathing has traditionally been practised. It was soon realised that compliance with the limit values would entail considerable investment in sewage treatment plant once the

UK could no longer rely primarily upon dispersion and dilution. The vagueness of the definition of 'bathing water' allowed the UK government at first to adopt a distinctly minimalist approach, with such traditional seaside holiday resorts as Blackpool and Brighton somehow not appearing among the twenty-seven beaches designated (fewer than landlocked Luxembourg) by 1979. A reasoned opinion of the European Commission in 1980 and the threat of 'infringement procedures'[22] forced the Department of the Environment to embark upon the process by which 464 UK beaches came within the terms of the Directive in 1995. Notwithstanding the gradual overall improvement, Table 5.1 indicates that it is not difficult to identify beaches all along the coastline of the United Kingdom (the north-west of England being particularly poor) where breaches of the standard for coliforms (bacteria indicative of untreated sewage) still occur.

Table 5.1 Bathing water surveys, 1989–95

Region	1989		1991		1993		1995	
Northumbria	32[a]	(37)[b]	33	(36)	34	(26)	34	(3)
Yorkshire	22	(18)	22	(14)	22	(5)	22	(9)
Anglian	28	(18)	33	(12)	33	(15)	34	(12)
Thames	3	(0)	3	(33)	3	(0)	3	(0)
Southern	65	(31)	67	(33)	67	(13)	67	(7)
Wessex	38	(18)	39	(8)	42	(17)	42	(5)
South West	132	(14)	133	(21)	133	(20)	134	(5)
Welsh	48	(17)	51	(12)	51	(18)	56	(12)
North West	33	(67)	33	(70)	33	(61)	33	(55)
England and Wales	401	(24)	414	(25)	401	(21)	414	(11)
Scotland	23	(30)	23	(35)	23	(22)	23	(17)
Northern Ireland	16	(0)	16	(0)	16	(6)	16	(6)
United Kingdom	440	(24)	453	(24)	457	(20)	464	(11)

Source: Adapted from *Digest of Environmental Protection and Water Statistics no. 14, 1991* (HMSO, London, 1991) Table 3.1, 36; and *Digest of Environmental Statistics no. 18, 1996* (HMSO, London, 1996) Table 4.1, 85.

Notes:
[a] For each year and each region, the number of bathing waters identified under 76/160/EEC is given first.
[b] The number in parentheses then denotes the percentage of these which breached (i.e. <5 per cent of samples failing) the limit values for total and faecal coliforms.

Swimming in waters known to be in breach of the limit value invites the charge of *volenti non fit injuria*; nevertheless, it is not too fanciful to imagine claims for compensation being made by those who suffer ill-health after swimming in such waters. There are reports of private law actions being initiated[23] by water sports enthusiasts, who suffer recurring intestinal disorders and eye infections, alleged to be caused by exposure to bacteria in inadequately treated sewage discharged to coastal waters.

The 'floodgates' prospect – damages being sought by a multiplicity of people who became ill after swimming in contaminated coastal waters and bathing lakes – may not be as stark as that of a polluted atmosphere, but it is not entirely absent. Dispersion and tidal mixing of marine discharges are inescapable physical phenomena. Their effects make fatuous the idea of one beach being effectively immune from compensation claims, whilst swimming from another can lead to action against a water company for a stomach upset. The burden of proof, which may involve a contest between epidemiologists, remains with the plaintiff. Recent research appears to suggest that the risk of contracting diarrhoea and sore throat is independent of measured levels of coliforms (bacteria) and enteroviruses, for which limit values are specified in the 1976 directive, but is related to strains of virus to which the directive makes no reference.[24] In another study,[25] the concentration of faecal streptococci in bathing water was found to be a better predictor of the subsequent likelihood of gastroenteritis; the authors argue that streptococci concentrations should replace those of coliforms as the basis for EC bathing water standards which, as given in 76/160/EEC, 'have very little public health significance to coastal bathing water in temperate north-west Europe'.

The coliforms present in samples taken from Blackpool and Southport beaches were the subject of the more recent[26] of the two occasions on which the European Court has held the UK to have breached a Community obligation towards the environment. The ruling of the Court was primarily concerned with dismissing the UK's claim that the directive required member states to take 'all practicable steps' rather than 'all necessary steps' to achieve the limit levels within the specified ten-year limit. However one reviewer of this case has postulated *Francovich* actions by those who can point to damage suffered as a result of violations of the limit values. The suggestion that *Francovich* would not apply because the Bathing Water Directive might not confer rights upon individuals can be readily countered, this reviewer argues, by the European Court's assertion that its aims included 'the protection of public health'.[27] However, the ruling in *Morton*[28] (see Box 5.2), although not a *Francovich* action, reveals little grounds for optimism.

Urban waste water

If the Bathing Water Directive sets limit values, the Directive on Urban Waste Water[31] aims to assist their achievement by ensuring that domestic sewage normally receives secondary treatment before discharge to estuaries or coastal

BOX 5.2 R. v. National Rivers Authority, ex parte Morton

Non-compliance with the Bathing Water Directive was one of the grounds by which a Tenby lifesaver sought a judicial review of a decision by the National Rivers Authority to approve (under s.85 of the Water Resources Act 1991) Welsh Water's discharge of sewage into the sea at Tenby. The other ground was that 'the NRA improperly took into account, and regarded itself as circumscribed by, Welsh Water's investment budget which set out the amount of money allocated for environmental projects'. According to the applicant, the discharge consent should have been subject to a condition requiring disinfection of the sewage by ultra-violet light which could have killed bacteria and viruses.

When drawing up its 'Asset management plan', Welsh Water had received advice from NRA on its environmental priorities. According to Harrison, J., if the NRA, when considering individual applications for discharge consents, took account of the exigencies of the applicant's financial position, it did not thereby exceed the limits of the statutory [para. 2.5 of schedule 10 of the 1991 Act] discretion to attach 'such conditions as [it] may think fit . . . for minimising the polluting effects of the discharges on any controlled waters'.

The ruling on this first count is simply a recognition of the breadth of the discretion traditionally bestowed upon regulatory bodies concerned with environmental protection. The second count is a contention that the principle of sympathetic interpretation (see Chapter 2) makes that discretion subordinate to the obligation to assist the fulfilment of the objectives of any relevant European law. Directive 76/160/EEC contains mandatory limit values, implemented in English law via regulations[29] and guide values, which a direction of the Secretary of State urges the NRA to achieve. According to the applicant in this case, a number of statements made by the NRA during the process of consultation which preceded the consent, implied that the limit values for enteroviruses were not seen as binding:

> [t]he virus standards . . . are not considered to have any scientific basis and the government has indicated that they should not be considered.[30]

The evidence led Mr Justice Harrison to the belief that the UK government was not implementing minimum EC standards on enteroviruses. He found the NRA's attitude to the standards was no more than 'equivocal' but, since the burden of proof of the authority's unlawful disregard of the standards lay with the applicant, the challenge failed.

waters. Primary treatment involves use of settlement tanks to remove most solids; secondary treatment subjects the resulting effluent to biochemical processes which enable it to be discharged to the aquatic environment, where dispersion and other natural processes further reduce the concentrations of polluting substances. The Directive, and the regulations[32] implementing it in the UK, allows secondary treatment to be dispensed with where the discharge is into 'high natural dispersion areas' (HNDAs) and where little environmental benefit would result from the additional treatment. Thus discharges into coastal waters from centres with a population in the range 10,000–150,000, and discharges into estuaries from centres with a population in the range 2,000–10,000, may be subject to primary treatment alone provided the discharge, in either case, is into a designated HNDA.

Two bones of contention, with obvious cost implications, immediately present themselves: does the outfall pipe lie in an HNDA or not, and is it within an estuary or coastal waters? But what constitutes the boundary of an estuary? Objective criteria such as salinity and topography – claimed the applicants in a judicial review[33] successfully challenging the decision of the Secretary of State to draw the boundaries of the Severn and Humber Estuaries so as to exclude discharges in areas declared to HNDAs. The minister was conscious of the considerable savings in water companies' costs (by dispensing with secondary treatment) if these outfalls were deemed to lie in coastal waters. The local authorities involved, with their general responsibility for public health in their areas, were unmoved by these considerations. They argued that defining the boundaries with a view to the consequent cost implications was unlawful. In accepting this submission, Mr Justice Harrison was aware of similar arguments (on the legitimacy of economic factors in the designation of special protection areas under the Birds Directive) which were then the subject of a referral to the European Court of Justice from the House of Lords[34] (see Chapter 8). (Fortunately, the opinion of the Advocate General, and the subsequent ruling of the European Court, was consistent with Harrison, J.'s decision made some six weeks earlier.)

Before concluding this section it is necessary to make passing reference to a recent case[35] in which an NGO attempted to force a local authority to take action to remedy some of the effects of discharging untreated sewage to the sea. A representative of Surfers Against Sewage sought a judicial review of the failure of Carrick Borough Council to take appropriate action against South West Water plc in respect of the statutory nuisance caused by the offensive 'sewage-related debris' (especially condoms and sanitary towels) washed up on a Cornish beach from that company's outfalls. It was not suggested that the deposits were 'prejudicial to health'; but they satisfied the basic common law test of public nuisance: they materially affected the comfort and quality of life of a number of citizens. Once nuisance is established, as it was in this instance, the duty to serve an abatement notice has long been mandatory under statutory nuisance law.[36]

This case is worthy of note not simply because of the applicant's success but because it made no reference to European law. Insofar as one can judge, standing was not an issue; it was reliant neither upon quantitative standards nor upon epidemiological or clinical evidence. With this type of nuisance, cause and effect tend to be both obvious and simultaneous; action in nuisance therefore remains a very effective remedy for a class of environmental ills (noise especially). And, as this case has shown, there is a role for private individuals (and NGOs) even in statutory nuisance. But it was a case which posed no challenge from existing environmental law nor, in terms of its empirical content, did it raise issues which could not have been understood by the parties in any of the venerable tort cases of the nineteenth century. To succeed in his action, the applicant needed to enjoy a right; but it was a *public law* right applied to a classic but depressingly persistent environmental problem.

Drinking water

The threat which agriculture can pose for controlled waters was recognised in the Control of Pollution Act 1974 where 'good agricultural practice' could constitute a defence (immunity) in criminal proceedings against polluters.[37] The high biological oxygen demand associated with effluent from silage clamps and intensive pig-rearing units can cause serious pollution of surface waters. In regard to drinking water sources, the principal threat lies in the nitrate, contained in both natural and artificial fertiliser, spread by dairy and arable farmers in the pursuit of higher yields per hectare. Notwithstanding the violation of the polluter pays principle, farmers in 'nitrate sensitive areas' are paid a subsidy to abstain from the use of natural and artificial fertiliser[38] (for further discussion see Chapter 8). The practice, from which the farmers are paid to abstain, contributes to the pollution of inland waters and, like the sewage contamination of coastal waters, has caused the United Kingdom to be in breach of a water quality directive.

The first occasion in which the European Court judged the United Kingdom to have failed to honour an environmental obligation followed Article 169 proceedings[39] in respect of breaches of the Drinking Water Directive.[40] The European Commission claimed, *inter alia*, that nitrate levels exceeded the 50mg/litre limit value in twenty-eight water supply zones (mostly in East Anglia). In rejecting the UK defence based upon the domestic notion of 'practicability', the European Court held that the duty to comply was absolute (even though the directive made provision for derogations in certain circumstances). In the aftermath of the ECJ ruling, Friends of the Earth applied in 1994 for a judicial review of certain decisions of the Secretary of State for the Environment concerning measures taken by two recently privatised water companies (Thames Water Utilities Ltd and Anglian Water Services Ltd) to reduce nitrate levels in their areas.[41]

Section 18 of the Water Industry Act 1991 requires the Secretary of State to issue an enforcement order in the event of a water company breaching any of its obligations. But if the defaulting company gives 'an undertaking to take all such steps as it appears . . . appropriate for . . . the securing or facilitating compliance',[42] then the minister is no longer obliged to make an enforcement order.

BOX 5.3 R. v. Secretary of State for the Environment, ex parte Friends of the Earth[43]

In the Court of Appeal, Friends of the Earth argued that securing compliance with the ECJ ruling by rectifying the breaches of the Directive justified curtailing, by passing legislation if necessary, the normal procedures for land acquisition and planning permission. The use of 'special development orders'[44] could avoid delays caused by the obligations, to advertise the proposals and to consult certain third parties, entailed in normal procedures by which any remedial works would receive planning consent (see Chapter 3). In dismissing this contention, Balcombe, L. J. could identify

> no requirement of Community law which requires the Secretary of State to ignore, or override, the provisions of domestic law, in particular where those provisions protect the rights of third parties which may include rights protected by the European Convention on the Protection of Human Rights and Fundamental Freedoms.[45]

It is questionable whether the public law rights of consultation – which might be curtailed by the use of a special development order – fall within the scope of the European Convention, although interference in the normal procedures of compulsory purchase of land could entail violations. However, it is difficult to take issue with the main thrust of Balcombe, L. J.'s conclusions. It is surely pushing the doctrines of supremacy and sympathetic interpretation too far to argue that a national executive must assume what almost amount to 'emergency powers' in order to honour any Community obligation.

The implications of the possible direct effect of Directive 80/778/EEC were taken up by Lord Justice Roch. He was aware that the European Court had clearly stated that the Groundwater Directive (see above) created rights for the benefits of individuals and environmental groups such as Friends of the Earth.[46] He accepted that the Drinking Water Directive could be treated similarly.

Member states enjoy a certain discretion in the way they implement directives into national law. However, it is now accepted that an administrative circular, whether published or not, is not an appropriate means of implementing directives which confer rights or obligations on individuals.[47] The key issue, for the transposition of all (not simply environmental) directives, is that in order to secure full implementation of directives in law and not only in fact, Member States must establish a specific legal framework in the area in question.[48]

That legal framework would be inadequate if individuals could not ascertain the full extent of any rights conferred upon them by the original directive. He was prepared to assume that 80/68/EEC created rights for the benefit of individuals but the attribution of rights was not 'free from difficulty'; there remained 'the question of the precise rights and duties the Directive creates, and which individuals are to have the benefit and the burden of those rights and duties'.[49] Having raised the question, he chose not to elaborate it; but like Lord Justice Balcombe, he did 'not understand European Law as requiring domestic law to confer on individuals enforceable rights which are completely uncircumscribed'.[50]

There is, as Balcombe, LJ. observed, 'nothing in the Directive [80/778/EEC, or for that matter 80/68/EEC on groundwater] which requires a member state to confer a specific right of action for damages under domestic law upon the individual'.[51] The point here is that there is no need for a 'specific right of action' provided that there exists a general one which can enable individuals to claim damages when any rights deriving from Community law are infringed.

According to Roch, LJ., both the European Commission and the Court had considered the Water Industry Act 1991 and judged it to be an appropriate means of compliance with the Directive. When given the opportunity, neither institution had questioned the apparent loss of a right of action where an undertaking, rather than an enforcement notice, applied. This difference, he pointed out, must be attributed not to the Secretary of State, but to the legislature which drafted and passed the statute. (Had this appeal occurred after the 1996 ruling in *Factortame*,[52] where the ECJ defined the liability of a member state when its legislation conflicts with Community law, he might have developed this point at more length.)

When the regional water authorities were privatised the Secretary of State accepted numerous 'undertakings'. Nevertheless, the continuation of numerous breaches of the Directive (in respect of nitrate concentrations) meant, according to Friends of the Earth, that he was not entitled to allow this less strict procedure.

Only passing reference was made to the rights of individuals at the unsuccessful application in the High Court. Late in the proceedings, counsel for Friends of the Earth explained that the breach of an enforcement order confers a statutory right of action (s.22 of the 1991 Act) on individuals suffering damage; the breach of an undertaking offers no such remedy and therefore the Secretary of State, by choosing the latter course, had denied individuals the possible enjoyment of this right. This issue does not appear to have been influential in the decision of Schiemann, J. to reject the application, although at the subsequent appeal (see Box 5.3) the question of individual rights was considered at more length. Their Lordships were clearly aware of the ECJ attitudes to rights conferred by environmental directives, especially 80/68/EEC protecting groundwater. However, these considerations were insufficient to persuade their Lordships to overturn the ruling at first instance: that the Secretary of State's obligation to comply with the ECJ ruling (in the Article 169 proceedings) was not compromised by his acceptance of the undertakings.

This case might lend weight to the argument that a directive should clearly set out the particular obligations which fall upon member states when in default[53] of any its provisions. For our present purposes, it is necessary to observe that, in *Friends of the Earth* as in *Morton*, the recognition of a directly effective provision of a directive was not sufficient to enable the applicants to secure their objective in a domestic court. The question arises as to whether an action in *Francovich*, by a water consumer from one of the affected areas, might have been more successful.

Francovich liability and the Drinking Water Directive

Francovich demands a remedy in national courts for damage causally linked to infringements of Community rights contained in directives, whether or not capable of direct effect. This particular remedy is indifferent to any immunities (such as the special protection afforded to 'undertakings') specified in national law. However the plaintiff's task – persuading an English judge that the three conditions for *Francovich* liability are satisfied – remains considerable. The notion that directives, which assist the 'protection of public health', thereby confer rights upon individuals was established in Community case law well before it was quoted in the German groundwater case.[54] That notion may seem no less relevant to an interpretation of the Drinking Water Directive; but it does not force an English judge to come to a particular conclusion in any action which might follow violation of that Directive. As we have seen, these rights have been inferred by the European Court, which considers them to be inherent in the Treaty of Rome. It does not require an unusually sceptical occupant of Queen's Bench to point to the absence of any explicit reference to rights in the text of any particular directive as sufficient grounds for dismissing an action relying upon direct effect or *Francovich*. Irrespective of the legal complexities, the plaintiff's case is far more likely to fall over the third condition of liability in *Francovich*: establishing a causal connection between damage and the breach of the member state's obligation in regard to drinking water quality.

Table 5.2 selects seven parameters and gives data on the numbers of contraventions (of the implementing regulations) in the 1990s. Whilst a consistent decline is apparent in most cases, progress towards total compliance is perhaps not as rapid as Friends of the Earth seemed to be demanding. Therefore it is not difficult to hypothesize examples of damage, stemming from violations of 80/778/EEC as with the Bathing Water Directive, for which redress in *Francovich* might conceivably be sought:

1   Dietary nitrate, when reduced to nitrite, can combine with haemoglobin in red blood cells and reduce the oxygen-carrying capacity of this compound. Children under twelve months of age are particularly vulnerable to methaemoglobinaemia (or 'blue baby' syndrome). However, the disorder has not been reported in children drinking mains piped water, except where bacterial contamination has also been present. The reduced reliance upon well and spring water in rural areas has effectively eradicated this potentially fatal condition; there have been no cases reported in the UK since 1972.[55]

2   Nitrates can react with certain compounds in food to produce nitrites and N-nitroso compounds, some of which are carcinogenic. Since the issue was first raised in the 1970s, epidemiological opinion has converged upon a consensus that the risk of stomach cancer from high dietary nitrate is minimal. A successful civil action, whether under European or UK law, by a cancer victim in a high-nitrate area seems very unlikely.

3   It is possible to imagine specialist companies in the food industry, or perhaps in pharmaceutical manufacture, for whom it was necessary to install and operate costly plant to denitrify a public water supply which exceeded the statutory limit in regard to nitrate. Action in *Francovich* to recover economic loss might then be possible.

4   The Commission's first Article 169 action against the United Kingdom was concerned with lead as well as nitrate (discussed above). However, the complaint in respect of lead pollution of drinking water, in seventeen supply zones serving some 52,000 people in Scotland, was not upheld by the European Court.[56] The argument hinged upon an interpretation of the obligation to take 'suitable measures' where lead concentrations of $100\mu g$ per litre are exceeded frequently. Irrespective of the technicalities of this case, there is no question that many UK citizens (thousands if not tens of thousands) continue to consume water with a lead concentration an order of magnitude greater than that now recommended by the World Health Organisation ($10\mu g$ per litre).[57] There is no dispute that this problem arises predominantly from contamination from lead pipes (and lead in the solder used in joining copper pipes) rather than, as with nitrate, an 'environmental' source such as lead in sewage sludge spread on land. The water companies can reduce plumbosolvency by reducing the acidity and softness of their water prior to distribution; but, according to Macrory,[58] they still have a duty of care in regard to the quality of water which issues from the taps of their customers.

Table 5.2 Drinking water quality in England and Wales, 1990–5

	1990	1991	1992	1993	1994	1995	(1995)
Iron	751[a]	799	802	671	614	618	(84)[b]
Lead	593	661	549	538	506	463	(59)
Aluminium	258	225	163	140	74	92	(11)
Nitrate	79	94	94	52	45	28	(1)
Trihalomethanes	83	127	135	116	58	37	(4)
Total pesticides	331	376	345	307	278	243	(29)
Coliforms	231	135	85	29	14	16	(6)
Total zones	2,536	2,577	2,598	2,576	2,552	2,471	

Source: Adapted from Department of the Environment *Drinking Water 1995: A Report by the Chief Inspector, Drinking Water Inspectorate* (1996, HMSO) and two preceding reports in this series.

Notes:
[a] Seven of the eighteen parameters given a prescribed concentration value (PCV) in the regulations (see ref. n40) have been selected; for each year and for each parameter, the number of zones in which the PCV was exceeded (even if only by a trivial amount) is given.
[b] For each parameter, the number in parentheses denotes the percentage of the total zones which, for at least part of 1995, were covered by an 'undertaking' (ref. n42) in respect of that parameter (irrespective of whether or not contraventions occurred in those zones in 1995).

In 1983, the Royal Commission on Environmental Pollution, claimed that:

> For most people in the UK, food and drink form the major pathway for lead uptake, but there is considerable uncertainty as to the relative contributions of the several sources of lead to this pathway.[59]

Since the publication of this very influential report, considerable effort has been made to reduce lead exposure from all pathways: leadfree petrol now constitutes 60 per cent of the market; and the removal of domestic lead piping has been encouraged by government grants. Nevertheless lead levels in drinking water exceed permitted concentrations in 20 per cent of supply zones in England and Wales in 1995 (see Table 5.2) and 12 per cent in Scotland. It is not too fanciful therefore to imagine *Francovich* actions in respect of inadequate implementation of Directive 80/778/EEC. Parents in an affected supply zone could argue, as in *Budden* (see Chapter 1), that their children's intellectual development has been impaired. It is unlikely that a future case would once again hang upon an apprehension of a 'constitutional anomaly' involving 'the sovereignty of [the Westminster] Parliament'.[60] It is even more unlikely that our hypothetical judge would hold, as Megaw, LJ. did in 1980, that the presence of lead in water meant that injury to the plaintiffs had to be 'assumed'.[61] In the light of current scientific understanding, a detailed scrutiny of the neurotoxicology of lead compounds would be inevitable.

Drinking water – a commodity?

The deeper one penetrates into the interconnected labyrinths of individual rights under national and Community law, the easier it is to lose sight of the extent to which water has become a 'commodity'.

The supply of wholesome water formed, from the earliest Public Health Acts of the last century, one of the definitive roles of local government in the UK. Cholera, dysentery and typhoid now very rarely threaten the wholesomeness of the water supply in developed countries. Some forty people died and many more were ill in an outbreak of typhoid in Croydon in 1937. In a test case, the father of one of the survivors successfully argued that the water undertaker was liable, not only for the breach of its statutory duty to him (as the owner of the property to which water for domestic use was supplied), but also in common law to his daughter (and anyone else) who might foreseeably be harmed by the undertaker's negligent actions.[62] Sporadic outbreaks of acute health effects following bacterial contamination of public water supplies have not been entirely eradicated; chronic exposure to low-level chemical pollution now generates more concern among environmental groups.[63]

When analysing FOE's judicial review of nitrates in drinking water, Purdue[64] suggested that individuals could sue for damages for the original failure to supply wholesome water if negligence could be established. Negligence was alleged in the Camelford case in 1988 – the contamination of drinking water in Camelford (a small town in Cornwall) when twenty tonnes of aluminium sulphate[65] were accidentally discharged into the supply system – which resulted in an out-of-court settlement of damages to a number of people suffering chronic health problems following the incident.

Some commentators[66] have argued that the Consumer Protection Act 1987[67] is wide enough to apply to water. Camelford attracted the environmental label, but with or without negligence it fits equally well within a 'consumer protection' tradition. Is there a sense in which Camelford's water supply system, wherein South West Water's duty of care was allegedly breached, is more 'environmental' than Stevenson's bottling plant upon which the hapless snail trespassed?[68] Neither involves a common property resource, and in neither case is the polluting matter dispersed within an environmental medium (namely natural water, land or the atmosphere). Although the victims' conditions were less serious, Camelford is in many respects comparable with the 'toxic oil syndrome' (which caused more than 350 deaths and affected some 20,000 people in Spain in 1981–2) to which the 'environmental' label does not appear to have been attached.[69]

In April 1994, dispersion in the River Severn proved insufficient to reduce the concentration of certain organic compounds below the odour and taste thresholds of water consumers. The substances originated in a solvent recycling business at

Wem (a small town in Shropshire); they entered[70] the sewer and passed undetected through the sewage works to the river and travelled eighty miles to Severn Trent Water's treatment works in Worcester and then into the supply network. Following a recommendation of the Drinking Water Inspectorate, the Secretary of State took action against Severn Trent Water for supplying 'water unfit for human consumption'.[71] The water company pleaded guilty and was fined £15,000 on each of three specimen charges and was ordered to pay costs of £67,000. These sums were exceeded by the *ex-gratia* compensation (£25 to each of the 60,000 households involved) paid in respect of the inconvenience incurred by the interruption in supplies. Public health was not an issue in this case, which was the first in which the phrase 'unfit for human consumption' has been judicially considered. There was no recourse to epidemiological evidence; the water was unfit simply by virtue of the consumers' reactions to its taste and odour.

Our concentration upon the quality of the water emerging from UK taps has been at the expense of a discussion of the question of a 'right' to be supplied with any water, irrespective of its chemical and biological contamination. Where no new mains are required, any owner of a property is entitled to be connected to the existing main for the supply of water for domestic purposes;[72] a parallel right exists for connection to the sewer and the discharge of domestic sewage (but not trade effluent).[73] But the continued rights of connection are conditional upon payment of the appropriate water charges. Disconnection was the ultimate sanction for non-payment of what, before 1989, were 'water rates' (effectively a property tax which, for tenants, was usually paid and then recouped in the rent by the landlord). Disconnection has obvious consequences in terms of personal hygiene and sanitation; these problems might involve the environmental health and welfare departments of the local authority. In the days when it was usual for the local authority to be the water undertaker for its area, overall interest might be best served by disconnecting only in the last resort and only in the event of wilful refusal to pay rather than true indigence. The existence of these other (health and welfare) agencies of the state with statutory responsibility for such families means that disconnection need not automatically amount to a violation of the right to water, which, as argued at the beginning of this chapter, is implicit in the right to life. Nevertheless, the European Court of Human Rights might choose to differ were a test case brought before it.

A study published in 1995[74] suggested that 'water debt' had increased by a factor of nine in the preceding five years. Water debt varied between water companies (5–13 per cent of households); as did the incidence of disconnection (0–36 per 1,000 households) but the overall rate was now declining from its peak of 12,500 in 1993–4. Interviews with a sample of defaulters suggested that, notwithstanding the 1992 'privatisation' of the regional water authorities into the water companies, there remains a view that 'water should be free' combined with a resentment at the high salaries paid to the executives of the water companies.

As with other utilities, privatisation of water has necessitated the setting-up of a new body, the Office of Water Services[75] (OFWAT) to exercise some residual control over the private-sector successors to the regional water authorities. Given the potential for abusing the 'natural monopoly' status of water, OFWAT's principal role lies in overseeing the operation of the mechanism which sets the charges for water services. Until the 'metering' of domestic supplies becomes more common and water charges more closely reflect consumption, then water perhaps will continue to be seen as a service rather than a commodity.

Conclusions

Does the pollution's eighty-mile journey by river make the Worcester case more 'environmental' than Camelford, where the contaminant was confined to the water supply network? Had biochemical activity in the river increased the toxicity of the original contaminants, then we would be forced to answer 'yes'. In this case, the river's role seems merely that of an incidental transport medium (along with the sewer and the water mains). Pollution of aquifers by nitrate in fertiliser depends upon physical processes: dissolution of nitrate and groundwater movement. The resulting contamination of potable supplies may fall within the scope of consumer protection law, but its origin merits the environmental label in a way that contamination, whether from lead pipes or from aluminium sulphate negligently emptied into a tank arising at the man-made end of the supply system, does not.

At common law, ownership of the foreshore is vested in the Crown. Economists might describe bathing from the foreshore as a public good which is non-rival in consumption. Thus there is a sense in which bathing beaches represent a 'common property resource' and any rights associated with the use of that resource attract the designation 'environmental'. A right to bathe in clean water would undoubtedly be infringed by pollution from sewage. But inherent difficulties in exercising that right cause us to doubt its existence. Where a beach is privately owned (e.g. a holiday camp) and bathing is part of a service which is bought, then consumer protection legislation or the law of contract might provide remedies for those subsequently suffering the effects of water-borne pathogens.

The enormous amount of money currently (and perhaps belatedly) being devoted to the infrastructure of water services in the UK might be interpreted as a recognition of a moral right to clean water, whether to drink or as a source of recreation. Yet our search for a legal right, which demands the 'environmental' label because no other seems appropriate, has not been fruitful. Tort remains a means of securing redress for those endowed with riparian and cognate property rights involving water; it is too early to assess the extent to which the ruling in *Cambridge Water Company v. Eastern Counties Leather plc*[76] and the partial erosion of strict liability, reduces its effectiveness.

Given the proliferation of directives related to the aquatic environment and their often incomplete implementation, opportunities now exist for individuals and, more to the point, environmental groups to use remedies established by European case law to expedite the Community's aim for clean water. The European Court has on occasions imputed individual rights to directives concerned with water pollution, especially those which protect public health. But as we have seen, it still requires the active support of the national judiciary to back these rights with effective remedies. In contrast, consumer rights have a longer history and are more readily recognised in English law. For that reason, we are inclined to conclude that, if there is a 'right to *drink* clean water', it is more likely to be found within consumer protection law than in the direct effect of directives concerned with water pollution.

6

RADIOLOGICAL PROTECTION

Introduction

In Chapter 2 it was claimed that a firm foundation for the European Community's environmental legislation dates from the Single European Act of 1987. However, in regard to radiological protection, UK policy has been subordinate to European obligations since accession to the Euratom Treaty (along with the other, better known Treaty of Rome) in 1973. In fact, the effects of ionising radiation from man-made sources (especially nuclear installations and radioactive waste) constitutes a specialised area of environmental management wherein UK freedom of action has been circumscribed by supra-national obligations since 1957, when the United Nations set up its International Atomic Energy Agency. Throughout this period, policy has been based upon 'recommendations'[1] prepared and published by the International Commission on Radiological Protection – a body of recognised experts in the fields of radiation biology, dosimetry, medicine, etc. It is difficult to cite another area of environmental management which is subject to a comparable degree of regulation or possesses as full a complement of treaties, statutes, regulations, advisory bodies and enforcement agencies.

With the breakdown in the wartime Anglo-American collaboration over atomic weapons research and the passing by the US Congress of the McMahon Act of 1946, the export of fissile material (i.e. plutonium and enriched uranium) as well as the exchange of information on nuclear research and technology was suspended. But successive British governments deemed the possession of an 'independent nuclear deterrent' to be politically imperative; in consequence the two Windscale piles (one of which caught fire in 1957) and the first chemical plant to separate the plutonium created were built. The world's 'first nuclear power station', commissioned in 1956 on the adjacent Calder site, was (and still is) a second-generation weapons-grade plutonium source. Subsequently the reprocessing of spent fuel from civil reactors became Windscale's (later renamed Sellafield) publicly acknowledged *raison d'être*.

Half a century later it is easy to forget that in the early days of the Cold War, perceptions of the strategic importance of the nuclear deterrent were such that safety and environmental impacts of the operations at Windscale (as at the equivalent US sites, Hanford and Oak Ridge) were at best a secondary consideration. Close connections between the UK civil and military nuclear programmes meant that public discussion of the environmental impact of both tended to fall foul of 'national security' considerations. This factor, combined with Whitehall's predisposition to secrecy, means that the nuclear field seems an unlikely place in which to extend our search for instances of the successful exercise of environmental rights. Nevertheless, those that are available are particularly rewarding. They span the range of private and public law remedies encountered in earlier chapters. Being concerned with the impact of radiation and radioactive waste, they relate to one of the key environmental issues of recent years.

An awareness of the scale of the damage which even a small accident at a nuclear plant might cause prompted legislation clarifying the extent of civil liability. One of the few actions brought under this heading has resulted in an unusually informative exposition of the limitations of epidemiological evidence in tort cases generally. Given their wider significance, the cases of *Hope and Reay*[2] are discussed at length in the first part of this chapter.

This is followed by an examination of some of the public law actions which have questioned the adequacy of the regulatory controls which, anticipating the risks posed by nuclear installations, seek to minimise the frequency of accidents and the permitted discharges of radioactivity. The commissioning of the new thermal oxide reprocessing plant (THORP) at Sellafield involved the discharge of mildly radioactive effluent to the Irish Sea. Greenpeace's challenge[3] of the Secretary of State's authorisation of these trace-active discharges resulted in a much publicised concession on *locus standi* with implications for other environmental interest groups. Examination of the evidence presented in this case reveals the extent to which the ICRP recommendations, on which most states with a civil nuclear programme base their regulatory regimes, is founded upon a utilitarian ethic indifferent to individual rights.

Statutory compensation for radiological damage

Of the various duties imposed upon a nuclear site licensee by the Nuclear Installations Act 1965, those contained in s.7(1) are of most relevance here:

to secure that

(a) No such occurrence involving nuclear matter . . . causes injury to any person or damage to any property of any person other than the licensee, being injury or damage arising out of or resulting from the radioactive properties . . . of that nuclear matter; and

(b) no ionising radiations emitted during the period of the licensee's responsibility –

(i) from anything caused or suffered by the licensee to be on the site which is not nuclear matter; or

(ii)   from any waste discharged (in whatever form) on or from the site,

cause injury to any person or damage to any property of any person other than the licensee.

Thus the licensee is strictly liable for injury to persons or property for 'occurrences' involving nuclear matter (with liability limited to £20 million for any one occurrence).[4] If 'no such occurrence . . . and . . . no ionising radiations . . . cause injury to any person' is interpreted literally, then it could be argued that no licence could ever be lawfully issued. There is no 'safe' level of radiation (see below); and some exposure, both of employees and of third parties, to radiation is inevitable in the operation of any nuclear installation.

This argument was rehearsed in 1987 in an abortive attempt by Friends of the Earth to secure a judicial review[5] of the decision of the Secretary of State for Energy to approve the construction of Sizewell B (the first pressurised water reactor but the last of any design to be commissioned in the UK). After dismissing the application on a procedural point, Gibson, L. J. opined in passing that Parliament could not have intended 'to prohibit the licensing of a nuclear power station unless it could be shown that no such occurrence could occur'.[6] In other words, the Court of Appeal recognised that it is scientifically meaningless to suggest that nuclear plant can be hermetically sealed units from which the emission of radiation and the escape of radioactivity is impossible.

Any claim based upon an alleged breach of the duty imposed by s.7(1)(a) hinges upon the existence of an 'occurrence involving nuclear matter'. If a massive release of radioactivity were ever to arise, and if workers or members of the public were to succumb to radiation sickness, then there is no doubt that this paragraph (together with all the associated provisions of the 1965 Act) would apply. But an illumination of the term 'occurrence' which recognises that the hazards posed by major nuclear installations form a spectrum – with catastrophes of Chernobyl proportions lying at one extreme, Three Mile Island in the middle, and minor departures from normal operations at the other extreme – is nowhere to be found in the 1965 Act (or any of its predecessors).

That incidents involving the escape of radioactivity into the atmosphere, and other accidents involving high doses (in excess of annual or quarterly limits) to individual workers have arisen during fifty years of nuclear fuel reprocessing at the Sellafield site in Cumbria, is undeniable. Moreover, the recognition of the possibility of such incidents is implicit in the numerous conditions of the Sellafield site licence which seek to minimise their frequency of occurrence. It was the accidental discharge of radioactivity into the sea, in breach of such a condition, which led to British Nuclear Fuels' (BNFL) conviction[7] on indictment and fines totalling £7,500 in 1983. However, none of the civil cases which we discuss in this chapter was concerned with the consequences of accidents.

Merlin v. British Nuclear Fuels plc

The Merlin family owned a house in Ravenglass, some six miles south of Sellafield. Their concern at the environmental impact of the reprocessing plant had been raised by the 1977 public inquiry into BNFL's proposal for a new plant (THORP) to reprocess thermal oxide fuel. They gave a member of a local environmental pressure group (Network for Nuclear Concern) a sample of house dust, collected in their vacuum cleaner, which was later analysed by Professor Edward Radford. The 11 picoCuries per gram (0.4Bq gm⁻¹) of alpha activity were, this distinguished American epidemiologist was later to claim, of grave concern to the health of the Merlin family. In April 1983 the Merlins assisted Yorkshire Television in making what was to become a controversial TV documentary *Windscale – the Nuclear Dustbin*. One sequence in this programme showed Mrs Merlin handing over other vacuum cleaner bags of dust for analysis. This publicity did not assist the Merlins in their attempt to sell their property.

The Merlins subsequently brought an action under s.12 of the Nuclear Installations Act 1965 in respect of economic loss to their property by radioactivity discharged via the pipeline from BNFL's plant to the Irish Sea and which then

> by the turbulence of the sea, found its way back to the coastline and became deposited in the mud of the Ravenglass estuary and thence, by the action of the wind and the carriage of the sediment on the feet of the plaintiffs, their family and their dogs, into their house.[8]

Aware of the risk that this contamination posed to the health of their children, the Merlins decided to move, and they put their house up for sale. The eventually agreed price of £35,000 represented, they claimed, a tortious loss attributable to potential buyers' apprehension of the effects of the now-widely known radioactive contamination.

The defendants did not dispute that the radioactive matter originated from their reprocessing plant. Their dismissal of the claim rested upon two points: the 1965 Act was never intended to offer remedies for purely economic loss; and it provides for compensation in respect of *proven* personal injury or damage to property; in other words, the Act is silent on risk or harm yet to be realised.

The first point was resolved by reference to the 1963 Vienna Convention on Civil Liability for Nuclear Damage.[9] According to Mr Justice Gatehouse, Article I of the Convention does not include economic loss within its definition of damage to property. It could however come within the meaning of 'any other loss or damage . . . if and to the extent that the law of the competent court so provides'.[10] Close inspection of the text of the 1965 Act nowhere suggests that Parliament had intended to extend the notion of damage to property beyond a commonsense understanding of the phrase.

Although it was Mr Justice Gatehouse's treatment of this first point which effectively demolished the plaintiffs' case, it was his incidental observations on the second which are of most relevance to this chapter. Without referring to *Re Friends of the Earth*,[11] he came to a similar conclusion:

> It is in the nature of nuclear installations that there will be some additional radionuclides present in the houses of the local population, over and above the naturally occurring radionuclides to which every one of us is continually exposed. If the mere presence of this additional source is enough to constitute damage under section 7 [of the 1965 Act], the result would inevitably be that the defendants were indeed in breach of their statutory duty every day to possibly thousands of citizens, each of whom have a claim for compensation.[12]

The force of the 'floodgates' argument – the prospect of a multiplicity of actions by individuals each claiming breach of statutory duty – may indeed dictate that authorised discharges under normal operating conditions can never be the subject of section 7 compensation.

In the only reported successful action[13] under this section of the 1965 Act, the common law reluctance to compensate for purely economic loss was maintained, but only by a generous interpretation of the physical damage involved. A pond containing radioactively contaminated water on the Ministry of Defence site at Aldermaston (where nuclear weapons are assembled) overflowed onto a small area of marshland in one corner of land owned by the plaintiff. The resulting contamination, although posing no threat to health, exceeded limits set out in the Radioactive Substances Act 1960. The problem only came to light in a survey by prospective purchasers of the land (including the much larger, uncontaminated area). Their subsequent withdrawal triggered the claim for damages. The defendants attempted to argue that in this case, as in *Merlin*, there was no 'damage' as there was no threat to human health. But the key difference in this case was the decision of the regulatory agency (HM Inspectorate of Pollution) to require decontamination so as to ensure compliance with the 1960 Act. For Mr Justice Carnworth, that decision indicated that physical damage had occurred and, in turn, economic losses resulting from it should be recoverable; following a complex calculation, he awarded the plaintiffs a sum in the region of £5 million.

In the event of an incident of Chernobyl proportions at a UK nuclear installation, victims of radiation sickness and the estate of any person who died from 'early effects' would have a clear claim. One must now assume that the large number of evacuees from houses, which required decontamination but were not otherwise damaged, would also receive statutory compensation. In terms of human mortality, the greatest contribution would arise in various forms of cancer in those exposed to high radiation but who survived the 'early effects' (e.g. radiation sickness). Insofar as compensation claims can be made up to thirty years after an 'occurrence', the 1965 Act recognises the existence of 'late effects'. However, the more distant in time and space an effect is from the occurrence, the more onerous becomes the burden of establishing the causal connection between them.

It is in every sense providential that the true effectiveness of the civil sections of the 1965 Act has yet to be subject to empirical test.

Hope v. *British Nuclear Fuels plc; Reay* v. *British Nuclear Fuels plc*

The Congenital Disabilities (Civil Liability) Act 1976 was intended to remove the confusion and uncertainty over the legal rights of unborn children, which had arisen during the protracted litigation undertaken on behalf of children born deformed as a result of their mothers having taken thalidomide during pregnancy. Section 3 of the 1976 Act defines the circumstances in which any child, born disadvantaged as a result of the exposure of either of its parents to ionising radiation from material on or emitted from a licensed nuclear installation, has *locus standi* to bring an action under the 1965 Act.[14]

Human genetic effects of radiation remain the subject of considerable dispute (see below) among scientists. In the absence of convincing evidence to the contrary, caution requires that radiological exposure of the reproductive organs must be assumed to give rise to some risk to the health of any subsequent offspring as well as to the exposed parent. This applies to both male and female parent, but we shall be concerned primarily with 'paternal pre-conceptual irradiation' (PPI).

The ensuing discussion of the cases of *Hope and Reay*[15] will be centred, not on legal points of interpretation as in *Merlin*,[16] but upon the consideration of the voluminous and substantive evidence which eventually led Mr Justice French to conclude that on the balance of probability, PPI, incurred via occupational exposure to ionising radiation at Sellafield, was not the cause of Dorothy Reay's fatal leukaemia or Vivien Hope's non-Hodgkin's lymphoma.

Hope and Reay (for all but the one question as to whether the two cancers have a common cause, they can be treated as one case) represents an opportunity to examine the relationship between law and science. This particular case demands attention for several reasons. Although undoubtedly adversarial, it went beyond the usual expert witness criticism of studies of the effects of allegedly toxic substances on animals in laboratories. The case was unusually open and accessible in that most of the evidence presented by both sides was already published in learned journals and was the subject of continuing and intense scrutiny within the scientific community. It adds an English perspective to a debate which tends to be dominated by American scholars and their closer acquaintance with the far greater amount of litigation in various US jurisdictions.

Within the literature, it is possible to find reports of cases in the United States in which large sums in compensation have been paid to individuals suffering health damage after exposure to various man-made pathogens in the environment. The readiness of lay juries to award damages on the basis of a superficial and highly selective exposition of the evidence of a cause–effect relationship has led some to argue that 'the legal system has determined that legal standards of proof need not be as rigorous as scientific standards'.[17] Others, in reviewing the same cases, conclude that 'the standards of proof in the courtroom are far more rigorous than those exercised in peer-reviewed medical journals – indeed those offering proof in court must do so under oath with the penalty of perjury, which does not happen in medical research'.[18]

But this emphasis upon a concept of 'proof' and upon the notion of a difference between legal (i.e. civil) and scientific standards of proof is somewhat misleading. Since Hume first encapsulated the problem, philosophers have continued to grapple with the question of causation. The fact that effect cannot be proven – in the sense that an axiom of geometry can be logically derived from certain initial premises – to follow from cause does not prevent us from making that connection constantly during each day of our lives. The exigencies of daily life demand that philosophical scepticism be suspended. The judge in his court and the scientist in his laboratory must take nothing 'for granted' but ultimately both must draw inferences from the information available. The degree of control available in certain branches of science in limiting the effect of extraneous factors enables those inferences to be made with a high level of confidence. The success of physicists and chemists in formulating robust theories must be attributed, in part, to their ability to limit the sources of variance in their closely controlled, laboratory-based studies. A judge can add his own questions to those of the rival advocates, he may exclude anything which violates the law of evidence but he cannot otherwise exercise control over the information before him and from which he (in the absence of a jury) must draw his verdict.

The activities of epidemiologists (or, for that matter, any social scientists) might represent an intermediate position. They study human populations and seek to identify associations between the incidence of disease and a suspected causal factor. They attempt, in the design of their studies and in the choice of the population, to minimise the possibility of any observed association being attributable to some unknown underlying (or confounding) variable. But that possibility can never be entirely eradicated, and the attribution of a causal status to any statistical association is made with extreme circumspection and following satisfaction of a set of stringent criteria. In the UK, those formulated by the eminent epidemiologist A. Bradford Hill[19] are most commonly used. Before examining these criteria, and their compatibility with the pragmatism of civil jurisprudence, it is necessary first to describe the origins of what became known as the 'Gardner hypothesis' and its principal rival, the 'Kinlen' hypothesis.

The Gardner and Kinlen hypotheses

The plaintiffs' reliance upon a genetic cause was based upon a hypothesis which had emerged from a 'case control' study[20] by the epidemiologist Professor Martin Gardner (two years before his untimely death in 1992). By demonstrating that juvenile leukaemia victims in West Cumbria tended to be concentrated among the offspring of employees (mostly males) at the Sellafield works who had incurred high radiation doses before conceiving these children, Gardner appeared to have broken the impasse which had arisen after studies of the populations around Sellafield and other nuclear sites had confirmed elevated rates of juvenile leukaemia. The excesses could not attributed to a direct (somatic) effect of 'environmental radiation', even when the highest estimates of the uptake of radioactivity in the respective localities were assumed. However, environmental radiation of the offspring might be the 'trigger' (or 'second hit') which induces cancer in those with the PPI-related propensity to the disease. This 'trigger' idea was to be raised by the plaintiffs at the trial, together with the possibility that the combination of the first and second hits could display 'synergy' (i.e. when two factors act together the resulting effect is greater than the sum of their effects in isolation).

In the aftermath of the notorious Yorkshire Television documentary of November 1983, the Minister of Health had commissioned an Independent Advisory Group (IAG) to study[21] an alleged leukaemia 'cluster' in the vicinity of the Sellafield nuclear fuel reprocessing plant. After concluding that the four cases of childhood lymphoid malignancy reported between 1968 and 1982 in Seascale (the electoral ward closest to Sellafield) could not be attributed to the radiological effects of that plant's discharges because these were calculated to give rise to, at the very most, 0.1 deaths, the IAG (of which Gardner was a member) was to incur scornful public criticism. When statistically significant excesses (over the numbers expected if the age-adjusted, national average rates applied) of juvenile leukaemia were subsequently identified around Dounreay,[22] the United Kingdom's other (but much smaller) reprocessing facility in the far north of Scotland, and in the locality of the nuclear weapons complex at Aldermaston[23] (some forty miles from Central London and where the environmental burden of radioactivity is many orders of magnitude less than at Sellafield) then the IAG's logic no longer seemed quite so perverse.

Gardner's paper excited the interest of the scientific community precisely because it appeared to offer a plausible explanation of a number of epidemiological studies which, considered individually, raised more questions than they answered. However, a series of studies[24-6] by Dr Leo Kinlen, Director of the Cancer Epidemiology Unit in Oxford, was soon to attract a similar interest.

If leukaemia were transmitted by some as-yet unidentified infective agent (e.g. a virus) and if a settled population of a remote area had developed a natural immunity to its own indigenous strain of that agent then, Kinlen argued, immigrants to that area would lack that immunity and excess leukaemias ('clusters') should become apparent following any large influx of population. Population mixing was characteristic of the construction and initial operation phases of remote nuclear installations like Sellafield and, even more so, Dounreay in the far north of Scotland during the 1950s. Comparable demographic changes occurred in less remote nuclear installations like Aldermaston in Berkshire, in areas designated for New Towns and in those areas in Scotland most affected by the exploitation of North Sea oil. Kinlen's claim to have identified leukaemia excesses in these non-nuclear sites did not go unnoticed by the defendant's counsel. Another of Kinlen's case-control studies,[27] of leukaemia among children living in Seascale but not necessarily born there, could not reproduce the association with PPI.

BOX 6.1 The Bradford Hill criteria in *Hope and Reay*

i Strength of the association

PLAINTIFFS The original Gardner study revealed a statistically significant (95 per cent probability) association between PPI and the excess leukaemias near Sellafield. Between publication of that study and the High Court hearing, further analysis (including two additional cases which had come to light) served to reinforce that association.

DEFENDANTS Without impugning the good faith and scientific integrity of its authors, the Gardner study was flawed.

   The association depended upon no more than four high-dose cases; this was extended to six with the additional data. But one of these – a nineteen year old male diagnosed while a student in Bristol – should have been excluded since the criteria for inclusion in the original study were that cases were to have been 'born and diagnosed' in West Cumbria. These criteria were chosen only after the start of a wider study of all cases diagnosed in West Cumbria wherever born (as the IAG[28] had recommended in 1983). Narrowing the criteria once a case-control study has commenced can introduce a bias.

   Irrespective of these methodological flaws, the Gardner study failed in its claim to explain the Seascale cluster. The spatial distribution of collective PPI dose (i.e. the aggregated dose received via occupational exposure at Sellafield by the fathers of Cumbrian-born children) showed no correlation with the spatial distribution of juvenile leukaemias.[29]

ii Biological gradient

DEFENDANTS No satisfactory evidence had been given which demonstrated a consistent tendency for the risk to increase with dose. The positive association was limited to the four cases in the high-dose (>100 milliSievert) category with no evidence of an elevated risk with lower exposures.

PLAINTIFFS The study gives a 'point estimate' of a PPI dose response (i.e. leukaemia incidence in progeny) relationship. The cause, if not PPI, of the excess leukaemias near Sellafield would have to 'mimic' that dose/response relationship.

iii Temporal relationship

PLAINTIFFS Response followed dose.
DEFENDANTS Not disputed.

iv Biological plausibility

PLAINTIFFS A number of expert witnesses demonstrated the existence of mechanisms which showed how radiation exposure of the male gonads could lead to a predisposition to leukaemia in offspring.

DEFENDANTS They were not arguing that it was biologically impossible for PPI to cause germline mutation which, with or without some synergistic factor, led to leukaemia in offspring. However if such an effect had given rise to the Seascale cluster, this would imply a far greater genetic component to leukaemia etiology than currently believed (recent studies suggest that the hereditable component of leukaemia is less than 5 per cent and even less for non-Hodgkin's lymphoma).

It would also imply that PPI, although having an unrivalled potency in inducing leukaemia, was not associated with other forms of cancer or malformations recognised to have a genetic component. If, as the plaintiffs had contended at this point, a 'synergistic factor' (or 'second hit') existed, then the combined effect would be far in excess of any comparable example of synergy known to science. There was also no evidence to suggest that this causal mechanism, if it existed, extended to non-Hodgkin's lymphoma.

v Consistency with similar studies

DEFENDANTS Two large-scale case-control studies – one modelled on Gardner's, of the offspring of nuclear workers in Ontario;[30] the other of all leukaemia cases in people under twenty-five in the years 1958–90 in Scotland[31] – failed to support Gardner's hypothesis on PPI.

The 'prospective cohort study'[32] of somatic and genetic effects in the survivors of the two atom bombs dropped on Japan in 1945 represents the largest and most closely scrutinised epidemiological study ever conducted. No significant excess of leukaemia has been identified in the children of the Hiroshima and Nagasaki survivors.

PLAINTIFFS If the results from studies of leukaemia incidence among the children of survivors of the A-bombs are correct then they are not 'quantitatively' consistent with Gardner. But Seascale is probably unique and possesses, in addition to PPI, three distinctive features: unusually high proportion of socioeconomic groups A and B; rural isolation; and high exposure to atmospheric radioactivity, which could be a synergistic factor (the 'second hit') with PPI.

DEFENDANTS Whatever this mysterious synergistic factor may or may not be, there is clearly enough of it in the Japanese environment to provide the 'second hit' necessary to complement the 'first hit', *viz.* the radiation which the A-bomb cohort survived in 1945.

vi Experimental corroboration

FRENCH, J. Of little relevance in this case, since there have been no laboratory experiments conducted upon live humans.

vii Coherence

Discussion which could be said to fall within this criterion was subsumed within others, particularly 'consistency' and 'biological plausibility'.

viii Analogy

Both parties were agreed that this criterion was inherently of little importance in this case.

ix Specificity

PLAINTIFFS Radiation has a 'scatter gun' effect upon cells; it is not surprising therefore that PPI caused a variety (acute and chronic, lymphoid and myeloid) of leukaemias at Seascale.

FRENCH, J. If, as the Defendants claimed, the absence of specificity weighs against the Gardner hypothesis, then it must act similarly in regard to the rival Kinlen hypothesis. This criterion was 'not very useful in this case because radiation is known to cause a variety of diseases, somatically at least'.[33]

In Box 6.1 we summarise, under Bradford Hill's eight criteria, the epidemiological evidence which appears from his summing-up to have been most decisive in Mr Justice French's ruling. It should be noted that discussion of the last criterion – specificity – appears to have been confined to what Susser[34] has labelled 'specificity in the effects', namely that cause X has, in the ideal, only one known effect Y. Demonstrating specificity in the effects, he argues, does little to strengthen a causal claim. In contrast 'specificity in the causes', namely that X is the sole known cause of effect Y (as exposure to asbestos dust is the sole known cause of mesothelioma) adds to the plausibility of such a claim. It can never be definitive, for it is always possible that empirical research will identify another equally convincing candidate for causal status. Since radiation, chemicals and viruses are individually associated with the leukaemia generation, none can claim 'specificity of cause'. However it appears to be only in this context of 'specificity' (albeit undifferentiated between causes and effects) that the Kinlen hypothesis was judged against the Bradford Hill criteria.

The Kinlen hypothesis, offering a common explanation for clusters in the vicinity of nuclear sites and elsewhere, has a greater 'predictive performance'[35] than Gardner's. Moreover, it has no need to resort to an as-yet unidentified 'synergistic factor'. One of the defendants' witnesses, Professor Sir Richard Doll (the doyen of UK epidemiologists, whose pioneering work with Julian Peto and Bradford Hill established the link between cigarette smoking and lung cancer) reported:

> I have previously concluded that the observation of an excess number of cases of leukaemia and [non-Hodgkin's lymphoma] in Seascale was not the result of PPI but could most reasonably be explained by a combination of chance and the effect of the sort of socio-demographic factors described by Kinlen and that the statistical association with [PPI] found by Gardner was probably due to a combination of chance and the *post hoc* selection of an atypical subgroup of young people who developed leukaemia in West Cumbria for concentrated study.[36]

This, one can assume, extinguished any remaining hopes for the plaintiffs, since Mr Justice French found it to be 'a conclusion which, on the evidence as a whole, seems to me no less plausible than the Gardner hypothesis'.[37]

At first sight, it might seem objectionable that a hypothesis which had not been subject to the rigours of the Bradford Hill criteria should be instrumental in demolishing one that had been. However, the onus of establishing their case (and hence the Gardner hypothesis) 'on the balance of probability' lay with the plaintiffs alone. The defendants were under no obligation to undermine that case, or to sow doubts in the mind of the judge, solely by reference to evidence which had survived equally stringent criteria. Moreover, we would argue that an unequal burden of proof does not represent any great disjunction between legal and scientific argument. It falls to the proponents of any 'new' scientific theory to demonstrate, via the range, rigour and the ingenuity of the tests it survives, why it represents an advance upon the status quo.

Following the ruling in *Hope and Reay*, the celebrated science journal 'Nature' published a commentary[38] by Sir Richard Doll and his co-workers in which he reiterated his reasons for considering the Gardner hypothesis to be wrong. This article may indicate a greater familiarity with scientific terminology than that shown by French, but in weighing the pros and cons of a large number of complex studies of different conditions among different populations, there is, I submit, little to choose between the learned judge[39] and the distinguished epidemiologist – they were engaged in what was essentially the same activity.

A contrary decision in *Hope and Reay* would have cleared the way for claims by other cancer victims among the offspring of the BNFL workforce. It might have persuaded BNFL to extend its non-statutory compensation scheme for its radiation workers so as to include provision for certain forms of cancer among their children. Thus it is possible to imagine the company incurring costs which, whilst not trivial, would hardly be crippling for an organisation of this size, given that the numbers involved – a point repeatedly made as a criticism of the Gardner study – are so small.

The Sellafield leukaemia cluster and its proximity to the UK reprocessing centre have been an environmental *cause célèbre* for more than a decade. It would be ironic if subsequent research into the Kinlen hypothesis were to corroborate the view that this and similar clusters arose through a social process (migration) rather than from an environmental factor. If the cancers of *Hope and Reay* were indeed the result of some infective agent introduced during the large-scale population mixing at the time of the construction of the original Windscale plant, who was at fault and who is to be sued? All the major construction sites (North Sea Oil facilities, Sellafield, the New Towns) and the large population movements (wartime evacuation, National Service) which according to Kinlen lend weight to his infection hypothesis, were not so much approved by Parliament (as with the lead-in-petrol level disputed in *Budden*,[40] see Box 1.2) but actually undertaken by agencies of the state, if not government departments. If the Kinlen hypothesis, after surviving a scrutiny as searching as that given to Gardner's, were to be widely accepted by the scientific community, claims for compensation by the estate of Dorothy Reay and others, whose leukaemia *may* have been the result of population mixing, would be no further advanced. Is the notion of fault applicable to the interconnected chains of events, decisions and responsibilities which leads to leukaemia (and perhaps other) clusters in such circumstances? Who, but the state, should compensate that sub-set of cancer victims whose condition might plausibly be said to have been the indirect consequence of construction work of national importance?

To conclude upon a further irony: the only group (other than employees at nuclear installations, veterans of the UK atomic bomb tests of the early 1950s and Blue Circle Industries[41]) to have received money from the Exchequer in compensation for damage attributable to nuclear radiation is comprised of those farmers in Cumbria, North Wales and Scotland whose sheep could not be sold following their uptake of radioactivity dispersed from the Chernobyl accident in April 1986. Restrictions on the movement of sheep were made under Part I of the Food and Environment Protection Act 1985, but this statute makes no provision for compensation. The UK government introduced an *ad hoc* scheme

> in recognition of the . . . failure of the USSR to acknowledge liability [under the Vienna Convention[42]] . . . and the real economic [cf. the ruling in *Merlin v. BNFL* above] difficulties faced by hill farmers.[43]

This compensation, which by October 1994 had amounted to over £11 million, was never debated in Parliament.

The International Commission on Radiological Protection

The term 'dose' (or more correctly 'dose equivalent') is a measure of the risk to human health (most notably in incurring cancer) resulting from exposure to ionising radiation. Since epidemiological studies have failed to identify a threshold dose below which adverse effects can be taken as zero, any exposure must be assumed to pose some risk, albeit small. The International Commission on Radiological Protection has developed three principles on dose limitation; the version which was current at the time (1977–90) of the events with which this chapter is primarily concerned reads:

(a) No practice shall be adopted unless its introduction produces a positive net benefit;

(b) all exposures shall be kept as low as reasonably achievable, economic and social factors being taken into account; and

(c) the dose equivalent to individuals shall not exceed the limits recommended for the appropriate circumstances by the Commission.[44]

The three ICRP principles formed the basis of a directive,[45] made under Article 30 of the (second) Treaty of Rome 1957 by which the EURATOM Community was instituted, laying down standards for the protection of both workers and third parties from the dangers of ionising radiation. The principles are also relevant to Article 37, under which plans for the disposal of radioactive waste from any nuclear installation must be submitted to the European Commission, who then refer these plans to a 'group of experts'.

HM Government's compliance with EURATOM obligations over radioactive waste management, and its acceptance of the ICRP framework as interpreted and elaborated by the National Radiological Protection Board (NRPB) in fulfilment of its statutory duty to advise the Secretary of State, was first explicitly stated in a 1982 white paper. This document also claimed that '[t]he requirements of the Directive are already largely met in the UK by existing legislation'.[46] At that time, the Radioactive Substances Act 1960 regulated the dose to the general public from radioactive wastes discharged to the environment, and regulations[47] which were issued under the authority of the Health and Safety at Work Act 1974 and which sought to protect employees from occupational exposure, were about to be amended.

The actual limit values specified in the directive for the dose uptake of workers (50 milliSievert) and the general public (5 milliSievert) need not particularly concern us. It is worth noting that this requirement, expressed as it is in quantitative terms and being precise, unconditional and concerned with human health, undoubtedly satisfies the criteria for 'direct effect' (see Chapter 2). In practice and following more recent ICRP advice,[48] lower dose limits are observed by relevant UK agencies. It is hard to imagine circumstances in which a member of the public, having exceeded the dose limit, would have recourse to 'direct effect' or action in

Francovich. Such a dose could only be incurred under the 'accident' conditions in which domestic legislation offers appropriate remedies. A worker at a nuclear installation could exceed the annual dose limit as a result of abnormal conditions, which need not necessarily amount to an emergency posing a threat to third parties (i.e. non-employees in the vicinity). Such circumstances, although rare, are not unknown in the UK. Again, one must assume that action in *Francovich* would apply. However, it is far more likely that the worker would be far more inclined to settle for the compensation awarded under some non-statutory scheme instituted by his employers (principally the nuclear power utilities).

Although there is no reported instance in which either workers or the public have sought compensation in the courts following exposure in excess of the requisite limit, this is not to deny that it is in these limit values of dose equivalent that the ICRP recommendations most closely impinge upon the type of 'rights' discussed in Chapter 2. The question remains whether there are less overt rights dimensions to be found in the application of the other two principles.

As low as reasonably achievable

The ICRP's elaboration of the phrase 'as low as reasonably achievable' (ALARA) is not unduly taxing in theory: it simply seeks to bring radiological exposure into the realm of conventional welfare economics (which, with Coasean bargaining, was encountered in Chapter 5). It represents an attempt to address the absence of a threshold in the dose/response relationship. Since *de minimis* does not apply, and a small but non-negligible risk will always attend the smallest dose, at what point can further dose reduction measures be dispensed with? That point is reached, and the ICRP's second principle 'is fulfilled at a value S* such that the increase in the cost of protection [X] per unit dose equivalent balances the reduction of detriment [Y] per unit dose equivalent', ie.

$$\frac{(dX)}{(dS)S*} = \frac{-(dY)}{(dS)S*.} \qquad {}^{49}$$

The quantity on the left-hand side is no more or less problematic than the marginal control curve associated with any other pollutant. However, the right-hand side of this equation is recognised as being in practice 'very difficult to quantify' but, ICRP continues, 'several estimates of the cost equivalent of a man Sievert [the unit of collective dose equivalent, i.e. a measure of the overall health detriment likely to be suffered by an exposed population] have been published and, with all their limitations, they provide possible quantitative inputs to the decision-making process'.[50]

Attempts to produce monetary evaluations of both mortality and morbidity attract criticisms from philosophers and economists alike. In the UK, responsibility for deriving the mathematical curve (representing the postulated variation in the monetary cost of collective dose with individual dose) has fallen upon the

National Radiological Protection Board. Their first attempt[51] appeared in 1981; the more sophisticated version,[52] published in 1986, is reproduced in Figure 6.1, which shows the factor which multiplies the base-line value (£3,000 per man Sievert) as a function of individual dose. This curve, when juxtaposed with a cost curve for dose reduction measures, enables designers and operators of nuclear plant to estimate levels of occupational and environmental dose which satisfy the ALARA criterion.

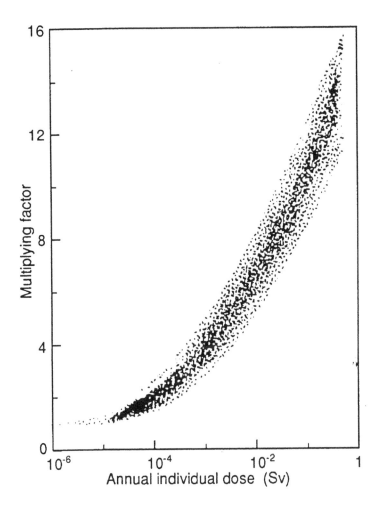

Figure 6.1 The monetary evaluation of collective dose

Source: National Radiological Protection Board.

The phrase 'as low as reasonably practicable' (ALARP), as developed over more than a century and a half of English factory legislation, is sometimes seen as synonymous with the more junior ALARA. The balance metaphor developed by the High Court half a century ago succinctly encapsulates 'risk/benefit analysis' (without recourse to the differential calculus used by ICRP; see above):

> 'Reasonably practicable' is a narrower term than 'physically possible', and implies that a computation must be made in which the quantum of risk is placed in one scale and the sacrifice involved in the measures necessary for averting the risk (whether in money, time or trouble) is placed in the other, and that, if it be shown that there is a gross disproportion between them – the risk being insignificant in relation to the sacrifice – the defendants discharge the onus upon them.[53]

The extent of the obligation on the operator of a nuclear installation to ensure safety was a central issue in the public inquiry[54] into the proposal by the Central Electricity Generating Board (CEGB) to construct a pressurised water reactor at Sizewell in Suffolk. It was originally intended that the nuclear licensing procedures, by which the Nuclear Installations Inspectorate (NII – a specialist division of the Health and Safety Executive concerned primarily with nuclear power stations and the Sellafield reprocessing plants) assessed the safety measures in the proposed design, should be complete before the Inquiry opened. This proved to be impossible; and in the 'rule 5' statement (see Chapter 3) made by the Secretary of State for Energy on the intended scope of the inquiry, there was, together with questions on long term energy options, waste management and general planning matters, a clause which indicated his desire to be informed of 'the safety features relevant to the design, construction and operation of the station and in particular the views of the NII as the licensing authority'.[55]

During this (record-breaking) 300-day Public Inquiry, safety was a very contentious issue. ALARA was soon agreed to be 'closely related' to ALARP, and attention became concentrated upon the meaning of the latter and the extent to which it had been observed in the design for what was the first of this type of reactor (using pressurised water as coolant) in the UK. In the evidence given at the Inquiry, it is possible to find only one substantive dispute[56] which was resolved by resort to an explicit cost-benefit calculation. This single example was concerned with minimisation of workers' exposure during normal operation. The major inconsistencies between the regulatory agencies and the nuclear industry, exposed during cross-examination and later reported Sir Frank Layfield,[57] lay in the application of ALARP to the minimisation of risk to both workers and the general public from abnormal conditions. Sir Frank later reported that 'ALARP rarely involved explicit cost-benefit analysis in the design and safety assessment of Sizewell B . . . [and] . . . the Inspectorate [NII] often appeared to have little regard to cost'.[58]

Given the lack of unanimity on approaches to safety, which Layfield's persistent questioning as well as that from the various anti-nuclear groups had exposed, it was, not surprisingly, the Secretary of State's consideration of the safety issues (including ALARP) which were the principal grounds of Friends of the Earth's legal challenge[59] to his decision (12 March 1987) to approve the applications.

After failing to persuade the Court that the civil provisions of the 1965 Act imposed a duty on the licensee of any nuclear installation to reduce emissions to zero (see above), FOE turned their attention to the duties imposed by the Health and Safety at Work Act 1974 in respect of risks to employees and to third parties. In particular, they argued that the Secretary of State had failed to address a particular requirement of the NII's *Safety Assessment Principles for Nuclear Reactors*:

> whether the frequency of any accident arising from a discrete fault sequence which could lead to a more serious release, and their summated frequency, had been made as remote as was reasonably practicable and whether all *reasonably practicable* steps had been taken to prevent such actions.[60]

The question of whether the Secretary of State did err in law in this matter is by no means straightforward. Had his decision letter made no reference to safety issues or to the extent to which they influenced his decision, then he would stand accused of inconsistency (ie. he failed to abide by his own Rule 5 intentions) as well as breaching the *Wednesbury* principle,[61] that when exercising a statutory discretion, relevant considerations must not be disregarded. But according to the Master of the Rolls:

> In any matter of this nature there was a wide spectrum of approach, any particular line of approach within that spectrum being justifiable and the choice of approach being that of the Secretary of State.[62]

There is no doubt that in the exercise of his power to determine such applications the Secretary of State does enjoy wide discretion. However, having originally declared that safety should form a key issue at the Inquiry, any conclusion relating to safety in his decision must take some cognizance of the relevant evidence contained in the report of that Inquiry. As Purdue *et al.*[63] have pointed out, he is not bound by his Inspector's conclusions and, in addition, faulty reasoning by the Inspector does not automatically invalidate the Secretary of State's concurrence with his recommendation. The thrust of FOE's argument was that, notwithstanding this wide discretion, the Secretary of State had to demonstrate why he was satisfied as to the reasonable practicability of the safety measures and that inherent in that demonstration was evidence that the marginal cost of further measures exceeded the marginal benefits. The protracted nature of the regulatory process by which developments like Sizewell B receive prior approval, and the growing sophistication of aids to decision-making, means that an obligation on the Secretary of State to consider the results of formal risk analyses does not seem to amount to an indefensible assault upon his discretion.

Sir Frank Layfield's concern at apparently inconsistent approaches to nuclear safety assessment adopted by different statutory agencies prompted the fourth of his fourteen recommendations:

> [HSE] should formulate and publish guidance on the tolerable levels of individual and social risk to workers and the public from nuclear power stations, recognising the limitations of present risk assessment techniques. In doing so the Executive should have regard to the benefits of nuclear power as well as the particular feature of the risks it creates.[64]

The result was the publication in 1988 of a document which became the basis for public exposition of HSE policy on the regulation of nuclear safety. It takes up Layfield's concern with 'tolerability', which implies 'a willingness to live with a risk to secure certain benefits and in the confidence that it is being properly controlled',[65] and contrasts it with the narrower concept of 'acceptability' with which risk analysts usually concern themselves.

The 'reasonably practicable' obligation imposed by statute on operators of nuclear plant is explained and interpreted using a simple but effective graphic device (see Figure 6.2). This reveals that tolerable risks, which pass the ALARP test, span a wide region between risks (such as liver cancer arising from exposure to vinyl chloride monomer) which cannot be outweighed by any associated benefits and those which are clearly negligible. Cost-benefit techniques are 'in principle' relevant in defining the boundaries of these regions. However, this view is hardly supported in Annex B of this document, which lists the conceptual and method-ological difficulties which have meant that 'the NII has so far found only a limited use for quantified [cost-benefit analysis] in aiding its decision-making about what reasonably practicable measures should be incorporated into nuclear power stations to reduce the probability or to mitigate the consequences of accidents'.[66]

An unprecedentedly detailed discussion of the NII's approach to the latter can be found in the latest version of *Safety Assessment Principles for Nuclear Reactors*.[67] The concept of tolerability is translated into a number of 'basic safety limits' for normal operations:

Example 1: 'No member of the public should receive in any year from all sources of radiation on the site a dose greater than . . . 1mSv'.[68]

and for accident conditions:

Example 2: 'The total predicted individual risk of death (early or delayed) to any worker on the plant attributable to doses of radiation from accidents should be less than 10^{-4} per year'.[69]

Coupled with each basic safety limit, there is a 'basic safety objective' (0.02mSv for Example 1; 10^{-6} per year for Example 2). This defines 'the point beyond which the

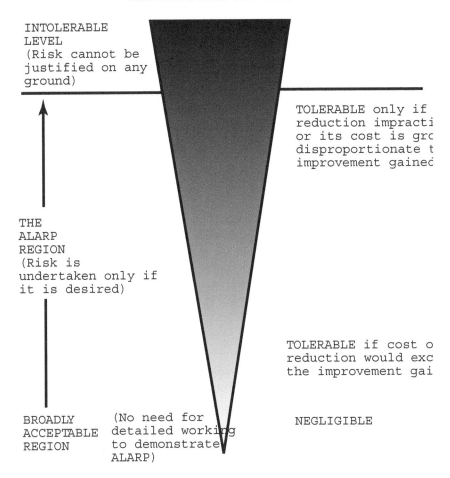

INTOLERABLE
LEVEL
(Risk cannot be
justified on any
ground)

TOLERABLE only if
reduction impracti
or its cost is grc
disproportionate t
improvement gainec

THE
ALARP
REGION
(Risk is
undertaken only if
it is desired)

TOLERABLE if cost o
reduction would exc
the improvement gai

BROADLY
ACCEPTABLE
REGION

(No need for
detailed working
to demonstrate
ALARP)

NEGLIGIBLE

Figure 6.2 The tolerability of risk
Source: Health and Safety Executive.

assessors need not seek further safety improvements from the licensee in his quest for ALARP'.[70] This latest interpretation of ALARP, I would argue, continues the tradition of placing greater emphasis upon the 'gross disproportion'[71] component than the balancing of marginal risks and benefits. Whether it has answered the criticisms voiced by Layfield is open to question but it is probably more intelligible to the layperson. The reliance upon quantitative 'basic safety objectives', as indicators of fulfilment of the ALARP criterion, when used in conjunction with the similarly quantitative basic safety limits, has the effect of blurring the distinction between the second and third of the ICRP recommendations. But the determination of the regulatory agencies to use quantitative limits as indicators of the achievement of ALARP/ALARA is perhaps more a reflection of pragmatism than an espousal of a rights-based approach.

The principle of 'justification'

In the formative years of ICRP, its first principle – the obligation to justify any practice entailing radiological exposure – related to relatively simple questions: if the mortality rate from side-effects of radiation doses incurred (by doctors and radiographers as well as patients) in the use of X-rays exceeded the lives saved by improved diagnosis, then the practice would be clearly indefensible. In applying the principle to an activity, like reprocessing or any other part of the nuclear fuel cycle, in which the costs and benefits are far more complex, the familiar limitations of utilitarianism soon become apparent. What weight should be given to, and how does one calculate, the increased risk of nuclear weapons proliferation? Can this risk be compared with the health detriment to present and subsequent generations (see Box 6.2)? The fact that ICRP's concept of justification entails quantification of net aggregated utility, whilst ALARA requires a comparison of marginals, does not make the calculation any easier. Technical considerations alone cannot explain the neglect of the principle.

BOX 6.2 **A brief digression on Sellafield's marine discharges**

In order to set the discussion of the need to 'justify' any radioactive discharge in a historical context, Figure 6.3 is included. Covering forty years up to 1995, it describes the pre-THORP era. The solid histogram denotes annual discharges of total alpha activity (measured in units of tera Becquerels (TBq) on the scale on the left-hand side) from the Sellafield reprocessing plant; the plot for total beta activity (not shown) has a somewhat different structure over this time period but not to the extent of refuting the points made below. The horizontal (broken) lines indicate the limit, on the annual discharge of total alpha activity, contained in a condition of the certificate of authorisation under the Radioactive Substances Act 1960 (and later the 1993 Act of the same title). The point of interest is the extent to which the authorised limit was raised in 1971 to meet the urgent need to increase discharges to sea. This was caused by a combination of production difficulties which meant that irradiated fuel from power stations spent longer in the storage ponds with a consequently greater corrosion of the magnox cladding. Equally, it should be noted that the authorised level has come down with the commissioning of ever more effective clean-up technology.

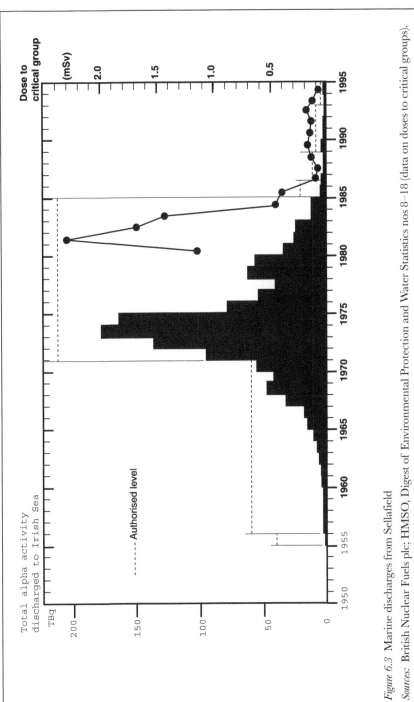

Figure 6.3 Marine discharges from Sellafield

Sources: British Nuclear Fuels plc; HMSO, Digest of Environmental Protection and Water Statistics nos 8–18 (data on doses to critical groups).

On the right-hand side of Figure 6.3, each black circle indicates the estimate of the annual dose (measured in milliSieverts) to the critical group for marine discharges, namely local fishermen whose diet includes fish and shellfish caught off the Cumbrian coast. Attempting to estimate doses for the 1970s is frustrated by a number of factors but different assumptions on consumption habits are the most important. These data on doses to the critical group have been extracted from successive annual volumes in the *Digest of Environmental Protection and Water Statistics* series. Some of these list two estimates: one uses the ICRP figure of 0.0001 for the gut transfer factor of plutonium, the other uses the NRPB-recommended figure of 0.0005. For consistency, the estimate based upon the lower figure is used throughout. Other sources of uncertainty mean that it is difficult to defend the individual points whilst having more confidence in the overall shape of the plot. The reason for superimposing these critical dose estimates is to demonstrate the 'time-lag' which exists between the marine discharges and their primary human impact. Absorption of certain radionuclides on marine sediments accounts for most of this lag which, from a cursory inspection of Figure 6.3, would appear to be less than a decade.

The time-lag associated with the terrestrial disposal of intermediate- and high-level waste will be measured in tens of thousands of years (following the completion of any underground repository). Although individual doses will be many orders of magnitude below than the lowest in Figure 6.3, they will be borne by far more than the two generations who have incurred what are in effect only the initial environmental costs (i.e. aggregated collective dose from Sellafield discharges to the Irish Sea, 1955–95) of Britain's nuclear (civil and military) programme. In view of the cancellation of the development of a commercial fast breeder reactor, the benefits – security through nuclear deterrence, electricity generation without dependence on fossil fuels – associated with reprocessing fuel used in that programme are similarly concentrated in the last four decades. This is merely to raise but one of the many issues to be analysed in depth before the obligation of 'justification' can be said to have been discharged.

There is a sense in which any action by a state which subscribes to the rule of law must be capable, at least in theory, of being justified. If an organ of the state were found to have acted arbitrarily, with clear procedural impropriety or in a manner which gave a particular individual, organisation or social group an advantage inconsistent with the agreed policy objectives, then any legal challenge of that action is likely to be upheld. Within UK jurisdiction, *Wednesbury*[72] tends to be cited in any case in which a decision of the executive was so unreasonable as to be unjustifiable. It is tempting therefore to argue that a specific legal obligation to justify that small minority of decisions of the executive which authorise exposure, of workers or the general public, to ionising radiation is superfluous. But just as ICRP defines ALARA to mean optimisation by more formal means than commonsense judgement or 'good engineering practice', so its interpretation of the justification principle implies a criterion which is more substantive than 'procedurally correct'. The current formulation of the first ICRP principle reads:

> No practice involving exposures to radiation should be adopted unless it produces sufficient benefit to the exposed individuals or to society to offset the detriment it causes.[73]

In an attempt to illuminate this principle, a two-stage model of decision-making is presented: in the first stage, the various options for achieving a given aim are identified; and in the second, a selection, which may entail replacing an existing practice with another, is then made. ICRP simply requires that exposure to radiation is not overlooked when other (possibly far greater) detriments of each option are assessed, along with the corresponding benefits. Moreover the Commission stresses that 'justification' is confined to the first stage: any option with a negative net benefit must be excluded from further consideration. As for stage two: 'To search for the best of all the available options is usually a task beyond the responsibility of radiological protection agencies'.[74]

The nature of the obligations entailed in the first ICRP principle was to be further examined in judicial reviews of ministerial decisions concerning radioactive discharges from the Sellafield reprocessing plant.

The judicial reviews of the THORP authorisations

In rejecting the application by Greenpeace for a judicial review of the authorisation of uranium commissioning (essentially a test procedure which did not entail full radioactive contamination of the plant), Mr Justice Otton[75] held, in September 1993, that THORP (BNFL's new complex at Sellafield for reprocessing spent oxide fuel from nuclear power plants such as Sizewell B) had already been 'justified in advance' by the extensive discussion at the 1977 Windscale Inquiry and by the parliamentary debate which (in line with the recommendation of Sir Roger Parker,[76] the Inspector at the Inquiry) approved the special development order[77] giving planning consent for THORP in 1978. According to Otton, J., these combined procedures 'inevitably [sic] took into account many more relevant issues including the benefit and disbenefit analysis to assess the economic value of the THORP installation'.[78] A few months later, in the judicial review into the legality of DOE and MAFF's authorisation of the discharges from a fully commissioned THORP, Mr Justice Potts specifically refuted this claim: 'Sir Roger Parker neither performed this exercise nor purported to do so. . . . Likewise with regard to the Parliamentary debates'.[79]

The fact that one High Court judge can assume, whilst another categorically refutes, that an assessment of the need for a development must 'inevitably' have formed part of the process by which it received planning approval, is eloquent evidence of the pervasive influence of merely 'procedural' justification. It should be added that Mr Justice Otton's error, if such it was, arose in the context of a subordinate argument for dismissing the application. In presenting his principal argument, he acknowledged the obligation (and its origins in the EURATOM Treaty) on authorising bodies to justify the practice (namely trace-active commissioning) giving rise to discharges, albeit that their radiological impact would be far smaller than those from a fully operational plant. It is therefore surprising that, in the ruling on Greenpeace's second application for a judicial review, Mr Justice Potts was required to dismiss claims by counsel for the Secretary of State for the Environment that no obligation to justify lay upon HM Government.

The Radioactive Substances Act of 1993 (which consolidated a number of amendments to the 1960 Act of the same title) makes no reference to justification or ALARA. Successive white papers[80, 81] have reiterated UK allegiance to the ICRP principles. More significantly, NRPB has, in fulfilment of its statutory duty to advise HM Government on various supra-national commitments[82] on radiological protection, formally endorsed the 1977 and 1990 recommendations of the ICRP. Counsel for the applicants chose to quote an extract from a guide to the 1960 Act, published in 1982 by the Department of the Environment,[83] which explained that, in accordance with NRPB advice, ICRP recommendations formed the basis of radioactive waste management in the UK. According to Potts, the obligation to justify stemmed, not from such sources ('Primary legislation is not to be construed by reference to general policy statements or departmental guidance')[84] but from the principle of Community law that national legislation should be interpreted in sympathy with the objectives of any relevant Community directive. EURATOM Directive 80/836 as amended, Article 6a of which reads:

> the various types of activity resulting in an exposure to ionizing radiation shall have been justified in advance by the advantages which they produce.[85]

was clearly relevant in this instance. Since 'interpretation of sections 13 and 16 [of the Radioactive Substances Act 1993] so as to accord with the Directive is entirely consistent with the purpose of the Act as spelt out in the Guide',[86] the Ministers had erred in law. Despite their error, the Ministers' earlier actions in regard to the authorisation were such, Potts averred, that the obligation to justify the discharges had in fact been met.

Potts' ruling on the first ICRP principle raises more questions than it answers. If 'the Ministers' approach to justification [could] not be faulted',[87] what exactly were the faults which they so assiduously avoided? The law of town and country planning, to make the obvious comparison, is replete with ministerial circulars and *dicta* on material considerations which, if not appropriately addressed, can lead to relief via statutory review. Few, if any, equivalents exist within the sparse law of radiological protection.

Mr Justice Potts' endorsement of the Ministers' approach relied heavily upon a detailed examination of their Decision Document[88] which had been written as if justification had been obligatory. Nothing in this document suggested that the Ministers were perverse or unreasonable, in the *Wednesbury* sense, in concluding that the THORP discharges were justified. On this point, Potts took pains to remind the Court of the legal principle that judicial review of administrative action is a supervisory and not an appellate jurisdiction: 'This Court cannot determine scientific and environmental arguments advanced to Ministers or resolve disputes of fact'.[89] However, the distinction between the legal and the substantive issues in a case of this nature is not entirely clear. In coming to the view that the Ministers had not neglected to consider any relevant matter, is not a substantive understanding of what constitutes an exhaustive set of pertinent issues required? Given the reasoning which persuaded Potts to endorse the Ministers' approach on the authorisation, it is hard to see why, in contrast to Mr Justice Otton, he was so insistent that the obligation of justification had not been discharged in the planning approval fifteen years earlier. A reading of Parker's report reveals a list of relevant matters (safety, proliferation, reprocessing versus storage, diversity of energy sources) no less exhaustive than that of the 1993 Decision Document. The disbenefits as well as the economic advantages to be gained from reprocessing both domestic and overseas arisings of oxide fuels were subject to cross-examination in public during the one hundred days of the Inquiry. It seems curious to argue that so protracted a planning determination did not subsume a justification obligation which, when dissected under the microscope of judicial review, is found to consist of nothing more substantive than *Wednesbury* reasonableness.

In 1978, Secretary of State Peter Shore rejected Parker's recommendation to approve THORP. This was necessary in order to satisfy the Commons' manifest desire for a debate (nominally over a special development order granting planning consent) into all the issues, not merely the conventional planning considerations, raised by the development. But Potts was as dismissive of the 1978 parliamentary debate as he was of the Parker Inquiry as a source of implicit compliance with the obligation to justify THORP. One suspects that few, if any, of the participants in the 1978 Commons debate could then have imagined that a judge in the High Court would later argue that theirs was not the last word on whether THORP was justified.

Discussion

In this chapter I have attempted to contrast private law actions for redress for damage alleged to have arisen from nuclear installations, with the public law actions in which third parties have challenged decisions by the executive in minimising the risk of such damage and in justifying that risk by reference to the associated benefits.

Irrespective of the attention paid to THORP at the planning stage and subsequently, it is without doubt the public inquiry into the Sizewell B pressurised water reactor which posterity will judge to be the high-water mark of the period from 1981 which O'Riordan[90] has labelled the 'age of public justification' of nuclear power in the UK. The fact that proceedings are conducted in public does not guarantee the openness which, together with fairness and impartiality, the Franks Committee[91] argued should characterise statutory public inquiries. These three principles, stemming largely from jurists' traditional understanding of procedural justice, must be distinguished from the deeper sense of *legitimacy*[92] which attaches to institutions of political decision-making whose fundamental rules are freely accepted by all participants. The opportunity for environmental groups publicly to challenge the technical evidence establishing the 'need' for new motorways, airports, toxic waste facilities, etc., must now be taken as a *sine qua non* of public acceptance of such developments. But it is necessary to remember that the much publicised concession on the standing of Greenpeace (to challenge the trace-active commissioning of THORP) was made at the discretion of Mr Justice Otton; the discretion of other judges to adopt a less liberal view (see Chapter 8) is undiminished. Similarly, the decision of Potts, J. concerning BNFL's costs in the subsequent review is not binding on those adjudicating similar challenges in the future. A more formal recognition of the role of environmental groups, and state funding of their participation in public inquiries, has yet to acquire a statutory basis.

With the Secretary of State's dismissal[93] (March 1997) of an appeal against refusal of planning consent for an underground test laboratory near Sellafield, the management of intermediate- and low-level radioactive waste, as set out in government white papers,[94] has suffered a setback. How severe and how lasting remain to be seen. Any alternative will have to survive an examination of the technical (especially hydrogeological) status of the proposed site no less stringent than that which took place in the sixty-day appeal against Cumbria County Council's earlier rejection.

Friends of the Earth, whose representative at the Inquiry had been instrumental in exposing technical flaws in the applicant's case, claim this decision to be 'the first time that the nuclear industry has lost a public inquiry in the UK'.[95] More importantly, it seems likely to be the last 'nuclear' inquiry for the foreseeable future. If the UK nuclear field has become fallow,[96] transport (see discussion of Twyford Down in Chapter 3) would seem to offer fertile ground for a continuation of the debate over the standing, and funding, of environmental groups. A repository for radioactive waste must resist water ingress for ten to one hundred thousand years. A decade's delay in the search for a suitable site might seem trivial in comparison. But it might provide an opportunity to consider the extent to which any process of justification – whether driven by rights or the calculus of utility – can take account of those many generations who will inherit the waste but enjoy none of the benefits stemming from the processes in which it arose.

7

THE EROSION OF PROPERTY RIGHTS

Introduction

The Fifth Amendment to the United States Constitution requires that any taking of private property for public use must be justly compensated. Similarly, Article 17 of the Universal Declaration of Human Rights holds that:

1 Everyone has the right to own property alone as well as in association with others.
2 No one shall be arbitrarily deprived of his property.[1]

English common law had adjudicated the rights and duties associated with the tenure of land for centuries before John Locke attempted the first modern philosophical defence of private property in land.[2] Expropriation of land and other property for the defence of the realm was once part of the royal prerogative. The standard procedure for the compulsory purchase by the Crown of private land now requires some order, contained in a statute, to be confirmed by a minister following a public inquiry at which the objections of persons with an interest can be heard; and any compensation is determined by an independent body (such as the Lands Tribunal in England and Wales). This obligation of consultation, right of appeal and payment of compensation mean that the standard procedure, if never exactly popular, is rarely seen as amounting to a violation of Article 17.2.

Chapter 1 has already considered – as an environmental right – an individual's right to take action in nuisance against those whose activities unlawfully interfere with the use and enjoyment of his land. The common law protected the right of an individual with an interest in land to enjoy it as he wished, without unreasonable interference from the occupants of adjacent land, and subject only to the obligations which derived from his tenure. Without denying the presumption in favour of an owner's freedom to enjoy his land, the exigencies of an industrial economy have necessitated gradual but inexorable growth in the circumstances which justify infringements of that freedom. The land use planning system and the state's assumption of the right to develop land were considered in Chapter 3. The subordination of an individual's rights of land use may be defended by reference to a wide range of communal goals (e.g. preservation of green belt, nature conservation, road networks). Given the aims described in Chapter 1, we shall concentrate our attention upon those justifications to which the 'environmental' label seems most appropriate.

In this and the following chapter, attention turns to various statutory regimes in which this historic right of landowners is subordinated to the needs of other interests. A clear distinction between the subject matters of these two chapters cannot be rigidly maintained. The present chapter focuses mainly upon controls exercised by local government over land whose future uses are circumscribed by its history (especially waste disposal and chemical contamination). In Chapter 8 the concern is with rural areas and with the protection of landscape and wildlife by means of controls enforced by specialist agencies of central government.

Civil liability and environmental damage

Marine pollution from the wrecks of the *Amoco Cadiz* on the Brittany coast in 1978 and the *Braer* off the Shetlands in 1993; the dioxins spread on land following an explosion at a factory in Seveso near Milan in 1976; the chemical pollution of the Rhine following a fire at the Sandoz warehouse in Basle (Switzerland) in 1986: these were some of the incidents with severe environmental impacts which have exercised the European Commission when it has considered the role of civil liability both in allocating responsibility for paying for the damage and also in ensuring the standards of risk-aversive behaviour which can minimise the rate of occurrence of catastrophic events. The Commission's 1993 Green Paper on remedying environmental damage[3] was concerned with the consequences of individual incidents like these and with the effects of long-term pollution which, whether permitted or not, cannot be described as accidental. The paper's remit extended to damage to the atmosphere, water and to land. However, attention in the UK has tended to become concentrated upon the issue of chemical contamination of soil by industrial activities over an extended period. Land contamination happened to be high on the Community's environmental agenda in the period immediately following Maastricht. Proponents of a strict and Community-wide regime therefore have an unprecedented array of principles – sustainability, precaution and high level of protection as well as the old stalwarts, polluter pays and prevention at source – to bolster their arguments. Opponents, however, can argue that contamination of land, unlike air and water pollution, does not cross national frontiers and therefore it is an area of environmental regulation which lends itself naturally to an application of the 'subsidiarity' principle.

If systems of civil liability for remedying environmental damage differed significantly among member states, distortions of competition and of the Single Market could result. However, the Commission assumed the existence of a 'basic and universal principle' – to be found, not in the Treaty of Rome, but in the civil laws of all member states – that 'a person should rectify damage that he causes'. The Commission also suggested that this 'legal principle is strongly related to two principles forming the basis of Community environmental policy',[4] namely prevention and polluter pays. To claim that a principle of this kind is universal is to believe that its key concepts can be found with substantially comparable roles in all jurisdictions. Given the manifold opportunities for semantic confusion, such a claim requires considerable confidence in the skills of the Commission's translators. As Steele[5] points out, 'rectify' cannot be taken as synonymous with 'repair'; liability in tort may require payment to make right the wrong done, but it does not necessarily require 'rectification' in the sense of a physical process which restores the *status quo ante* even though that might be possible.

The UK government asserted that the Green Paper did not reflect 'the differing legal traditions of Member States'.[6] And since the Green Paper included no example in which differing systems of civil liability could be shown to have led to distortions of competition, subsidiarity should apply: '[h]armonisation is not a goal in its own right and there is a need for a convincing case to justify action at Community level'.[7] Historical differences in land use – the UK's legacy of industrial contamination being far greater than that of Greece or Portugal – justified decision-making at national level. To sum up the UK's overall approach to this (and other) environmental problems, the Response reiterated a paragraph from the 1990 White Paper on the Environment:

> Action on the environment has to be proportionate to the costs involved and to the ability of those affected to pay them. So it is particularly important for Governments to adopt the most cost-effective instruments for controlling pollution and tackling environmental problems. And we need to ensure that we have a sensible order of priorities, acting first to tackle problems that could do most damage to human life or health and could do most damage to the environment now or in the future.[8]

In keeping with the Brundtland definition of 'sustainable development', further harm to health and the environment will be prevented or minimised; as for existing damage, HM Government aims to 'encourage remediation where practicable'.[9] This insistence upon cost effectiveness was later to be enshrined in the statutory controls (discussed below) over contaminated land which, at the time of this response, were still under consultation.

Save for this one proposed addition, the response considered the existing statutory controls and common law remedies to be adequate and appropriate to address the *remediation* of environmental damage. Before turning to this latest addition, it is useful for our purposes to examine some of the ideas emanating from Europe which the Department of Environment found unappealing.

A proposal to extend, to non-governmental organisations with general environmental objectives, certain powers which supplemented those of the regulatory bodies, although included in a Council of Europe Convention on environmental damage,[10] was not supported. Glossing over the contentious question of the standing of public interest groups, the response argued that the public regulatory bodies 'are accountable for their actions on behalf of the community as a whole';

and judicial review enabled 'interested persons and organisations to challenge the authorities' decisions in the Courts'.[11]

The reasons for opposing 'joint compensation funds' are not rehearsed at length. It is accepted that there may be occasions when it is more appropriate to spread the cost of clean-up across an industry rather than concentrating it upon a small number of firms which might thereby be bankrupted. But voluntary schemes allow some members to opt out in the hope of securing a short-term competitive advantage; in turn, compulsory schemes suffer from problems in ensuring efficiency.[12] This later view is supported by a reference to the US Superfund (see Box 7.1), but no further elaboration of the failings of this particular scheme is offered.

Since a superfund may be obliged to insure itself against catastrophic loss, it is constrained by the attitude of the insurance market to environmental risks. Insurance is identified as the key issue in the House of Lords Select Committee's report on the Green Paper. Their Lordships recognised the problems in securing cover for the unrealised environmental impact of past activities, and they confined their attention to liability for future pollution. Their understanding of experience in marine insurance[13] led them to conclude that a superfund, to which all potential polluters within a given industrial sector contributed, was necessary to cover for costs which commercial insurers could or would not meet:

> The insurance market even when more mature than at present, will not
> be able to cover all possible environmental liabilities. Financial cover
> beyond a set high level will need to be provided by such means as joint
> compensation schemes.[14]

In their response[15] to this view, DOE reiterated its opposition to obligatory compensation schemes on the grounds that they could result in remediation costs incurred through the carelessness of a minority of firms having to be borne by those whose environmental safeguards were far superior. In addition to this efficiency argument, the DOE pointed to (what were then draft) proposals for remediation of contaminated land in the Environment Bill. These, it was claimed, would remove uncertainties in the current liability regime and encourage a less diffident participation by commercial insurers.

When the Royal Commission considered the 'superfund' in 1985, it too found the concept less than compelling:

> There may be a case in either equity or economic efficiency for passing
> clean-up costs . . . on to the present customers of the polluting industry
> rather than to the general taxpayer, on the ground that these costs should
> properly have fallen on past customers; but we do not consider the case to
> be a strong one.[16]

The USA was then estimated to have as many as 20,000 abandoned hazardous waste sites, of which Love Canal[17] was perhaps the most notorious. There was no evidence that a problem of this scale existed in the UK where, the Royal Commission argued, effective controls over land use had been in existence for longer. After more than a decade's experience of the US Superfund's operation, it is the widespread resentment at the enormous transaction costs (well in excess of sums actually spent on clean-up) which makes the greatest impression. It is this political reaction, rather than differences in relative need, which ensures that it is unlikely to be replicated in the UK. The spiralling cost of superfund litigation encouraged the adoption of 'alternative dispute resolution' (ADR) methods by the Environment and Natural Resources Division of the US Department of Justice. The successful US experience[18] of arbitration (binding and non-binding), mediation, fact-finding, mini-trial and other alternatives to full, adversarial trial in court has not been lost on those responsible for the administration of civil justice[19] in the UK.

BOX 7.1 CERCLA and the Superfund

The primary purpose of the US Comprehensive Environmental Compensation and Liability Act of 1980 (CERCLA) is to empower the Federal Government (usually via the Environmental Protection Agency) to respond to releases of hazardous substances which pose a threat to public health or the environment. The costs of responding may be recovered from those subsequently identified as responsible for the release; but the absence of a clearly identifiable perpetrator does not preclude action. The responses range from 'immediate action' (subject to a $2 million maximum commitment from the fund) to forestall a release, or limit (for example) its contamination of water supplies, to 'remedial action' which offers a permanent reduction in the risks posed by the released substances.

The original fund of $1.6 billion proved woefully inadequate in the face of estimated total liabilities in the range $150–300 billion. In 1986 Congress voted a further $8.6 billion to the fund; this sum included $2.75 billion raised on a petroleum tax and $2.5 billion on a corporate environment tax on companies with an annual turnover in excess of $2 million; only $0.3 billion came from monies recovered from 'potentially responsible parties' (PRP) liable from clean-up operations. PRPs include the present and past owners of the site, those responsible for the disposal of hazardous substances and those responsible for its transport. Given the joint and several liability of PRPs, the costs of legal transactions between the various parties have become of the same order as the remediation costs, to a point which has caused considerable damage to the reputation of 'Superfund'.

Particular problems arose over the liabilities of banks and other institutions lending money for activities involving hazardous substances.

Under CERCLA, a holder of a security interest is normally exempt from liability, but exemption can be compromised by actions which imply some measure of 'control' over a hazardous site. *Fleet Factors Corporation*[20] had a security interest in a cloth printing facility which subsequently went bankrupt. In the ensuing sale of assets, Fleet is alleged to have given instructions to contractors engaged to remove unsold equipment and to clean away the now-empty factory. In January 1984 when the EPA inspected the site, they found 700 drums of toxic chemicals and forty-five lorry loads of asbestos-contaminated material. In the ensuing legal action to recover the $400,000 incurred in disposing of these materials, the Court of Appeals (Eleventh Circuit) held that Fleet's instructions to the contractors was a sufficient exercise of 'control' to cross the threshold of operator liability. Awareness of US experience, and this case in particular, prompted the House of Lords Select Committee to recommend that a 'lender's liability should be restricted to circumstances where he exercises effective control of the damaging activity'.[21]

This discussion of environmental liability, and of 'Superfund' in particular, has become concentrated upon hazardous substances on land, not because the concepts do not apply to other media (air and water), but because the low rate of terrestrial dispersion of pollutants means that the problems tend to involve a longer time scale. The normal procedures of compensation by strict and fault-based liability systems are frustrated, not so much by difficulties in isolating a culprit within a multiplicity of sources (as with air pollution, Chapter 4), but by the problem of identifying a guilty polluter in what may become a long list of successors in title[22] of polluted land. The greater the spatial separation and the interval in time, between cause and effect, the greater the chance that the identity of a polluter will be lost (or that he might become impecunious). The Superfund established by CERCLA has been described as 'a statutory attempt to mould a "common law" of hazardous waste cleanup'.[23] It is now necessary to consider the statutory system of regulating contaminated land, introduced into England and Wales by an administration which, unlike the European Commission (see Table 7.1), viewed the implementation of a complementary system of joint liability as unnecessary.

Table 7.1 Applicability of civil liability in instances of environmental damage

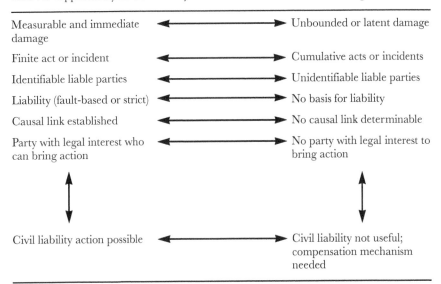

Measurable and immediate damage	⟷	Unbounded or latent damage
Finite act or incident	⟷	Cumulative acts or incidents
Identifiable liable parties	⟷	Unidentifiable liable parties
Liability (fault-based or strict)	⟷	No basis for liability
Causal link established	⟷	No causal link determinable
Party with legal interest who can bring action	⟷	No party with legal interest to bring action
Civil liability action possible	⟷	Civil liability not useful; compensation mechanism needed

Source: European Commission, *Green Paper on Remedying Environmental Damage* (see note 2)

Contaminated land

BOX 7.2 *Cambridge Water Company v. Eastern Counties Leathers plc*

Although Eastern Counties Leathers has been manufacturing fine leather at its Sawston tannery since 1879, it was only since the early 1950s that it has used organochlorine chemicals to remove grease from animal skins. In 1983 groundwater, abstracted by Cambridge Water Company from its borehole some 1.3 miles from the tannery, was found to contain trichloroethene and perchloroethene in concentrations in excess of the limit value set in the EC Directive on the quality of drinking water.[24] In moving their abstraction plant to a point where concentrations did not disqualify this aquifer as a source of potable supplies, CWC incurred a cost in excess of £900,000. Even though the storage of substantial chemicals on industrial premises could be said to constitute 'an almost classical example of non-natural use',[25] CWC's action for recovery of this cost and an injunction to prevent further contamination failed. After lengthy argument in which Lord Goff appeared to uphold the view that the rule in *Rylands v. Fletcher*[26] is an extension of the tort of nuisance, he concluded that foreseeability (which did not arise in this case) should be regarded as a prerequisite of liability under that rule.

There is a certain irony in the fact that the rule in *Rylands v. Fletcher*, which was invariably cited in the textbooks as a remedy for environmental damage, should be undermined by a case (see Box 7.2) whose 'hard' environmental credentials cannot be questioned. The organohalogen compounds found in the Cambridge aquifer were List I substances under a European directive,[27] the very purpose of which is to prevent contamination of groundwater resources. But it is perhaps premature to see the rule as moribund. Advances in the understanding of hydrogeology, continued concern over possible carcinogenesis by organohalogens, their persistence, and not least the continuing interest generated by *CWC v. ECL* ought to conspire to make a defence of unforeseeability far less convincing if invoked in comparable circumstances in the future. Foreseeability, like reasonableness, is not a static concept. Hindsight is both wise and unforgiving: once the effect of something has been seen, it is much harder on a subsequent occasion to claim that it was unforeseeable.

The consequences of ECL's past activities were now distinctly foreseeable and therefore, counsel for CWC argued at one point, they were liable for the *continuing* pollution of the aquifer. Lord Goff was unmoved by this argument. According to one commentator,[28] the adoption by the European Union of the Council of Europe's Convention on Civil Liability for Damages Resulting from Activities Dangerous to the Environment[29] would give Cambridge Water a statutory cause of action for current and future percolation of trichloroethene and perchloroethene (assuming these were covered by the Convention) from Eastern Counties Leather's land. But this, like other refinements of civil liability regimes, is contingent upon the identification of a defendant (and one who is not impecunious). The long history of industrialisation in the UK has resulted in contamination of groundwater and surface water by leachate from long-abandoned quarries, gasworks, mine workings and solid waste tips where the identification of the 'polluter' responsible is simply not possible.

Contamination of land is a 'material consideration' to be taken into account when determining planning consent for development of that land, and such sites may be identified in any appropriate development plan. Planning controls are of limited application in regard to *existing* land uses; the Environment Act 1995 promises a new regime for the 'remediation' of contaminated land which is independent of town planning and statutory nuisance law.

When first passed, the Environmental Protection Act 1990 contained no new powers concerned with the clean-up of contaminated land. In 1989 the House of Commons Environment Committee had recommended that *caveat emptor* should no longer apply insofar as the seller of land should inform any potential purchaser of any contamination. The 1990 Act imposed a duty (s.143) on every district council in England and Wales to prepare a public register of land in its area which had been put to one or more 'contaminative uses' (to have been specified in regulations). Given the costs of site investigation and subsequent laboratory analysis, it was decided that an entry in the register should record potential, rather than actual, contamination. One effect of this decision was that sites which had been restored could not be removed from the register. The absence of hard data drawn from field measurements did not allay the fears of prospective purchasers. Banks and other property interests became extremely apprehensive about this novel form of 'blight'; and their concern was influential in securing the repeal of this section before it had come into force.

The new regime, comprising Part IIA of the 1990 Act, does include a public register but now this is concerned with land which is actually contaminated. This term is to be understood, not in terms of the concentration of any toxic substance in soil, but in relation to conditions in which

(a)   significant harm is being caused or there is a significant possibility of such harm being caused; or

(b)   pollution of controlled waters is being, or is likely to be, caused.[30]

Further elaboration of this important definition is promised in guidance to be issued by the Secretary of State. In common usage, the term 'contaminated' usually implies the presence of a foreign substance, irrespective of any damage or harm. However, a draft[31] issued in May 1995 instructed a local authority to disregard, for the purposes of defining contaminated land, any harm or interference other than:

*   harm to human health of users or occupiers of the land in its current use or to health of current human users or occupiers of other land
*   harm to or interference with the ecosystems protected under the Wildlife and Countryside Act 1981 [and two EC Directives, see Chapter 8]
*   harm to property (including livestock, crops) in relation to the present use of the land or other land

Furthermore 'harm to human health' is to be understood to mean only 'death, serious injury or clinical toxicity'.

Every district council in England and Wales is to 'cause its area to be inspected from time to time'[32] in order to identify contaminated (as defined) sites. This survey is the necessary precursor to the process which the Department of Environment has continued to label 'remediation'. Following a three-month period of consultation with those involved, the local authority[33] serves a 'remediation notice' on the person (or persons) who caused or knowingly permitted the actions which led to the contamination and who, in keeping with the 'polluter pays' principle, must bear the cost of remediation. This of course assumes that original polluter can be identified within what might be a long succession of transfers of the land. Where no such 'polluter' can be found, then the notice is served upon the current owner or occupier.[34] The effect of the 'polluter pays' principle is, wherever possible, to anchor liability for clean-up with the original agent of the contamination. This applies even if the contamination was known to the purchaser and was reflected in the price. Moreover, statutory liability under the 1990 Act is indifferent to any rights or indemnities agreed privately between the parties.

151

Government policy, in as much as it can be inferred from draft guidance issued for consultation, suggests that the term 'contaminated', as the criterion for entry to the register, implies only that the land is not suitable for its *current* use;[35] and the definition is indifferent to any other to which it might be put in future. This policy will clearly minimise the liabilities and remediation costs of those who purchased land in ignorance of its chemical history. Whatever its intention, it is hard to see how this policy can be reconciled with the precautionary principle or indeed with the concept of sustainability. As with planning circulars, local authorities must 'have regard to' any published guidance when deciding the extent of decontamination measures to be specified in a remediation notice. They enjoy a measure of discretion and could, within the limits of *Wednesbury* reasonableness, require a standard of remediation which goes beyond that demanded by its current use.

Landfill

It is recognised that sites used for the disposal of household and industrial wastes will be prominent in the contaminated land surveys in most districts. Even if landfill (the term 'tip' should be reserved for sites at which illegal or uncontrolled disposal occurs) posed no threat of contamination, it would still be necessary to give it our attention since it remains the principal form of waste disposal in the UK. And waste disposal, or more precisely, the need to pursue more sustainable approaches to the management of waste, currently occupies centre-stage in UK environmental policy.

In theory that policy has been subject to the obligations imposed by the 1975 Framework Directive on Waste[36] but in practice the converse was true, since it is now recognised[37] that Part I of the Control of Pollution Act 1974 (in draft at the time of UK accession) was used as a template for the Directive. The basic aim of the Directive is that the 'competent authority' should license disposal operations so as to minimise the threat to public health and to the environment. Member states were also to encourage the reduction of waste and the recovery of materials and energy from waste materials. An attempt by an Italian environmental group to invoke the 'direct effect' of a provision of this Directive in order to secure a greater commitment to recycling met with a predictable lack of success.

An Italian decree which relied exclusively on landfill and failed to encourage recycling was considered to be in violation of the requirement on all member states to take

> the necessary measures to ensure that waste is disposed of without endangering human health and without harming the environment, and in particular: without risk to water, air, soil, and plants and animals, without causing a nuisance through noise or odours, without adversely affecting the countryside or places of special interest.[38]

But following an Article 177 referral, the European Court[39] held that the provision at issue set out the framework for action to be taken by the member states; it failed both the precise and unconditional tests and did not confer rights which individuals could rely upon in national courts. Although this is yet another example of an unsuccessful attempt to invoke the 'direct effect' of an environmental directive, it is difficult to dispute the suggestion[40] that this particular case was unlikely to resolve the uncertainties surrounding the extension of this doctrine to the environmental aspirations of the Community.

The collection and disposal of domestic refuse have been functions of English local government since the sanitary reform period of the last century. The regulation of all disposal routes, not simply landfills, became a local authority[41] responsibility under the Control of Pollution Act 1974. The move towards privatisation in the 1980s led to the separation of the local authorities' regulatory role from the operational responsibility (which most had assumed) for the landfills, incinerators and other facilities which received both domestic refuse collected by district councils and (on payment of a fee) waste from industry. It is easy to forget that the principal purpose of the Control of Pollution Act 1974 was to strengthen criminal sanctions against 'fly-tipping' and to encourage disposal at licensed sites, managed by qualified engineers, where technical controls could minimise nuisance and threats to the environment. The particular dangers to public health and to the environment posed by the 'fly-tipping' of hazardous chemicals necessitated additional controls. Under regulations[42] covering 'special wastes', a consignment note had to accompany such waste from its departure from its place of production, through any transfers, to its final disposal at an appropriately licensed site; and the waste disposal authority had to be notified at each stage. This regime of accountancy continues to underpin the current controls over special wastes;[43] it now extends to the wider class of controlled waste.

The 1990 Act added another weapon to the regulators' armoury: s.34 imposes a 'duty of care' on persons involved in the disposal of controlled waste to take all reasonable steps to ensure that they, or any others to whom the waste is transferred, can and do complete the disposal without committing any offence under the relevant sections of the 1990 Act. Guidance on the discharge of this duty is contained in a code of practice which the Secretary of State is obliged to prepare; the code is admissible in evidence in any court proceedings in regard to a breach of the duty of care. In one of the earliest prosecutions following a breach of this duty of care, a metal treatment company was fined £800 when tins of hazardous paint residues were found burning near ponds in Cleveland (NE England). Papers found near the fire led the waste regulatory authority to the company, who then admitted to having passed the waste to an itinerant scrap dealer.[44]

The incentive to fly-tip special waste lies not only in the higher charges paid at the disposal site, but also in the extra transport costs which may be involved in reaching a site licensed to accept it. When English county councils were waste disposal operators (as well as regulators) exhausted clay pits became highly prized since, with their natural impermeability, they provided a site from which 'leachate'

– toxic liquid waste formed by the microbiological breakdown of organic waste – could not escape and contaminate underground waters. In fact, the high level of biochemical reactivity created by household waste could be utilised in the breakdown of chemically hazardous waste from industrial sources. However, this practice of 'co-disposal' is being phased out in view of opposition from the European Commission. Methane gas is another product of the microbiological breakdown of organics and, under the right conditions, poses a threat of explosion[45] when allowed to accumulate in a confined space. It also makes a significant contribution to 'greenhouse' global warming.

Leachate contamination and methane generation are the principal risks associated with landfill. The recent policy shift away from landfill, driven primarily by the European Commission, stems more from a recognition that it is the least acceptable form of waste disposal in terms of sustainability. Not only does the practice itself proscribe certain uses of land following restoration, but the availability and relative cheapness of landfill acts as a disincentive to an earnest commitment to recycling of materials, recovery of energy and waste avoidance in manufacture. For that reason it occupies the lowest place in the hierarchy of preferred options given the National Waste Strategy for England and Wales.[46] It is possible to point to other, earlier measures which have been proposed to reduce UK dependence (a 20 per cent reduction over twenty years is the target) upon landfill:

- various schemes, under the 'non-fossil fuel obligation',[47] which seek to recover the energy potential in organic material;
- the publication in 1993 of the Report of the Royal Commission on Environmental Pollution, which concluded that high-temperature incineration was the 'best practicable environmental option' for the disposal of hazardous waste;
- the obligation (s.49 of the 1990 Act) on every district council to prepare a 'waste recycling plan' for its area;
- the target, set in *This Common Inheritance*, of 25 per cent of UK household waste to be recycled by the year 2000.

When the data on current disposal routes are examined, the enormity of the task of reducing reliance upon landfill, even for special waste (Table 7.2), becomes apparent. The capacity of the UK's four merchant incinerators is limited. The burning of wastes with high calorific value in cement kilns (see Chapter 3) is, with the consequent saving on fossil fuel, somewhat more 'sustainable' than incineration alone (and far more so than landfill), but it is still unacceptable if it consumes materials (e.g. industrial solvents) which could be treated and re-used. Even if he had the power[48] to do so, there has been no suggestion that the Secretary of State would use it to dictate the disposal route to be followed by all hazardous substances. The directive on hazardous waste incineration[49] advocates that 'any heat generated should be used as far as possible'. It is possible that authorisations for individual hazardous waste incinerators could be amended so as to include a capability of heat recovery; but it might be argued that such a requirement could conflict with their original *raison d'être*, namely the high-temperature destruction of toxic organic compounds.

Whilst environmentalists might see the inherent problems in lessening dependence upon landfill as justifying the resolute application of 'command and control' measures by the state, the UK's most recent response has been to apply a market mechanism.

As originally proposed,[50] the 'landfill tax' would have been analogous to value added tax in that it would have been proportional to the charge which landfill operators impose per tonne of waste (but unlike VAT, it would not have been deductible as an input tax on the waste producers). The *ad valorem* nature of the proposed tax incurred a lot of hostility: although ostensibly consistent with 'polluter pays', it would encourage waste producers to seek out the sites with the lowest costs regardless of their environmental standards. In the event it was a weight-based tax which eventually appeared in the Finance Act 1996: £2 per tonne of inactive (in the sense that it – building rubble is perhaps the best example – does not pollute groundwater or generate methane gas); £7 per tonne for other categories. Thus domestic waste and hazardous waste incur the same rate of tax when landfilled.

Part of the revenue raised by the landfill tax, estimated at around £400 million per year,[51] can be used to fund certain activities of 'environmental bodies'. An environmental body can be any legal entity whose objects can be found within a list of approved purposes, which includes research into recycling, the remediation of former landfills where ownership and liability is uncertain, and the restoration of any building of historic or architectural interest. In 1996, HM Commissioners

Table 7.2 Disposal and treatment routes for arisings of 'special waste' in England and Wales

	*1992—3**	*1993—4**
Landfill	1.490	0.931
Treated	0.553	0.619
Incinerated	0.162	0.197
Recycled	0.129	0.185
Stored	0.012	0.000
Total	2.346	1.932

Source: Adapted from Table 7.4(a) *Digest of Environmental Statistics no. 18, 1996* (1996, HMSO) 127.

Note:
* Million tonnes; the source data are rounded to the nearest thousand tonnes.

of Customs and Excise set up a regulatory body, Entrust, to enrol the bodies and ensure that any contributions they receive are spent on approved purposes. Landfill operators are able to claim a rebate of 90 per cent of their payments to enrolled bodies (up to a maximum of 20 per cent of their annual landfill tax liability); and it is estimated that £100 million per year could become available for approved purposes. A local authority may set up any environmental body but it cannot control it. Some concern has been expressed over the ethics of the practice of certain local authorities, when awarding contracts to landfill site operators for the disposal of municipal waste, to insert clauses requiring donations to be confined to projects which they, the local authorities, consider most deserving of support.[52]

It might be unduly cynical to dismiss this recently imposed levy on landfilled waste as a means of raising revenue without incurring widespread unpopularity. It is necessary to recall (see Table 7.3) that mineral extraction – the largest single source of solid waste, of which the great majority is simply deposited on land – is exempt from the tax. Moreover, the tax of £2 per tonne may not be high enough to persuade the construction industry to make real efforts to recycle demolition wastes (for example, as aggregates and in road construction). Nevertheless the landfill tax has been generally well received and could lead to a further 'greening' of fiscal instruments. A direct tax on aggregates, if it encouraged recycling, could reduce the environmental impacts of both the extraction and the ultimate disposal of these minerals.

Discussion

For the private individual, the Environment Act 1995 created no new 'rights' as such. According to one commentator, no new liabilities are created: the 1995 Act rationalises liabilities for contaminated land which already exist under statutory nuisance and water pollution law.[53] However, the obligation on each local authority to 'cause its area to be inspected from time to time' for contaminated land (as distinct from statutory nuisances) is undoubtedly an innovation and, if duly observed, will lead to the realisation of liabilities which might otherwise have remained notional. The decisions of Environment Agency, over contaminated land as well as air and water pollution, will no doubt occasionally be challenged by individuals and NGOs exercising their public law rights to seek judicial review. This process will be assisted by the right of access to registers of 'remediation declarations'. However, as I have argued in Chapters 1 and 2, to label purely procedural rights as 'environmental' adds little to our understanding.

The translation of the concept of 'duty of care' from the common law of negligence into the principal statutory controls over waste disposal was originally proposed by the Royal Commission on Environmental Pollution in 1985.[54] It is not insignificant that one of the more innovative provisions (s.34) of the Environmental Protection Act 1990 consisted of a duty, with no reference to any correlative right.

156

Table 7.3 Estimated annual waste arisings in the United Kingdom (Mtonnes)

Agriculture		80*
Sewage sludge		34*
Minerals:		
coal	17	
slate	7	
china clay	24	
quarries	32	
total minerals		80
Industrial:		
blast furnace slag	6	
power station ash	13	
other	50	
total industrial		69
Demolition and construction		70
Dredged spoils		36
Household		20
Commercial		15

Source: Adapted from Table 7.1, *Digest of Environmental Statistics no. 18, 1996* (1996, HMSO)
 125.

Note:
* Wet weight; the disposal on land of these two categories can be consistent with 'good husbandry' if
 not sustainability.

This chapter has been far more concerned with duties than with rights. Earlier chapters on air and water have been concerned largely with obligations owed by potential polluters to persons spatially separated from them. Given the persistence of terrestrial pollution, the obligation is more to persons, separated from us in time, who might still be adversely affected by a hazardous legacy of our industrial economy.

Silica is the most abundant compound in the Earth's crust. On a geological timescale, the effects of anthropogenic activities which require silica to be quarried in one place, to be transformed into bricks and concrete, and to end up eventually in a landfill in another are trivial in comparison to the natural processes of erosion and sedimentation. The next glaciation will change the landscape of Scotland far more dramatically than the loss of the Caledonian Forest, the Highland Clearances or any of the other human activities which created the 'natural' wildernesses which we now so earnestly seek to protect. It is legitimate to object to 'superquarries', such as that proposed at Lingerbay on the Isle of Harris,[55] on the grounds that it denies future generations one of the opportunities, which my contemporaries and forebears have enjoyed, to find inspiration in one part of the rugged coastline of the Western Isles. It is our grandchildren and generations of the *near* future who are most deprived by such development.

8

LANDSCAPES, SPECIES AND HABITATS

Introduction

For centuries trees, plants and animals, both wild and domesticated, were seen as part of the property of landowners. The statutes which make property rights subordinate to the aim of conserving animal and plant species, irrespective of their food or other instrumental values, are comparatively recent phenomena. The first statutory protection of certain species of sea birds dates from 1869 and was followed by gradual extension to other, terrestrial species. Its one million members make the Royal Society for the Protection of Birds the largest conservation charity in Europe. As the royal affiliation suggests, this body has always been part of the 'establishment'; its objectives have been enormously assisted by the endorsement and active participation of members of the political elite. Together with the Council for the Protection of Rural England, these bodies have exerted a powerful influence in the genesis of the large number of statutes which offer direct protection of landscape and of various species of plants and wild animals. However, this chapter's discussion of the role of state in wildlife conservation will be primarily concerned with the growing hierarchy of what must be labelled indirect controls, namely those which seek to protect the habitats of endangered species.

Discussion will be concentrated upon rural areas and upon the 'environmental' impacts of modern agriculture. Rachel Carson's warnings[1] on the consequences of pesticide use for bird populations did much to remove complacency on both sides of the Atlantic. In a later but similarly influential book,[2] Marion Shoard provoked concern with the visual impact on the British countryside of modern farming practices. Many of these practices were encouraged and funded by the Common Agricultural Policy (CAP) of the European Community. Before joining in the chorus of disapproval which invariably greets this policy, it is necessary to recall the extent to which our uplands and other much-loved landscapes – whose 'loss' is attributed to the CAP subsidies which encourage intensification of rural land use – were themselves the consequences of earlier agricultural practices.

Attempts to regulate what is often referred to as the 'natural' (as distinct from the 'built') environment predate UK accession to the Community by centuries. William Blake was not the first, nor Aldo Leopold[3] the last, to recognise the conflict between those with an economic interest in land and those who see the countryside (and its flora and fauna) as a source of intrinsic value. This chapter and the next examine the extent to which the legal protection accorded, directly and indirectly, to plants and animals reflects values other than the purely instrumental. Is it possible to extend the protection of intrinsic value; and is co-existence with the conventional view – nature as a resource to be exploited – achievable?

Protected areas

Disputes between ramblers and landowners anxious to protect their grouse moors culminated in the famous mass trespass on Kinder Scout in Derbyshire in 1932. The subsequent Access to Mountains Act 1939 was replaced by the far more extensive National Parks and Access to the Countryside Act of 1949.[4] This included a procedure by which an order, confirmed by the Secretary of State, designated an extensive area of land as a 'national park', wherein natural beauty could be preserved and enhanced and where public access[5] and open-air recreation were to be assisted. The 1949 Act makes it clear[6] that preservation extends to the park's flora, fauna, geological and physiographical features. All but one (Pembrokeshire Coast) of the ten National Parks in England and Wales are situated in upland areas. (Nearer sea level, the Norfolk Broads and the New Forest are now protected by specific statutory arrangements equivalent to those of national parks.) The forty Areas of Outstanding Natural Beauty contain a wider range of landscapes, with the chalk downlands of southern England being well represented.

Where an area of land needs to be managed in order to preserve or assist research into certain flora and fauna, that area may be designated as a 'nature reserve' under the 1949 Act.[7] The Wildlife and Countryside Act 1981 (the '1981 Act') has empowered the various conservancy councils[8] to confer the special status of 'national nature reserve'; and approximately half have been so designated. The Royal Society for the Protection of Birds (RSPB) has entered agreements with the conservancy councils to manage 129 reserves. However, it is the 'sites of special scientific interest'(SSSI) which constitute the most numerous category within Table 8.1 and, in terms of total area covered, are exceeded only by the 'environmentally sensitive areas' (see below).

The protection which the SSSI designation affords an area has been used to secure compliance with certain international obligations concerning wildlife. All terrestrial UK wetland sites of importance for waterfowl, designated under the Ramsar Convention, are SSSIs (as are the 'special protection areas'[9] and the 'special areas of conservation',[10] to be discussed at length below). The first SSSIs were set up under the 1949 Act, s.23 which obliged the conservancy council to inform the relevant local planning authorities of that designation. That provision was replaced by the more stringent requirements[11] of the 1981 Act: where a conservancy council considers land, by reason of any flora, fauna, or geological or physiographical features, merits the protection offered by SSSI status, it must notify the owners and occupiers of the land as well as the relevant local planning

Table 8.1 Protected areas in the United Kingdom

Status	No.	Area(km²)	(%)
National parks[a]	10	13,729	9
Areas of outstanding natural beauty[b]	49	24,070	15
National scenic areas[c]	40	10,020	13
National nature reserves	333	1,990	—
Local nature reserves[d]	487	250	—
Marine nature reserves	2	30	—
SSSI[d]	6,178	20,410	—
ASSI[e]	72	750	—
Special protection areas	104	3,270	—
'Ramsar' wetland sites	88	3,620	—
Environmentally sensitive areas	38	31,080	—

Source: Adapted from Tables 8.15–16, *Digest of Environmental Statistics no. 18*, (HMSO, London, 1996).

Notes:
[a]   England and Wales
[b]   England, Wales and Northern Ireland
[c]   Scotland
[d]   Great Britain
[e]   Areas of special scientific interest (Northern Ireland)

authority and the Secretary of State. The notice had to specify the features constituting the 'special interest' and the operations which it prohibits; and any landholder (i.e. owner or occupier) had three months in which to appeal against the proposed SSSI designation of any of his land.

Ministers are also empowered under the 1981 Act[12] to designate, by statutory instrument, areas of moorland or heath within national parks where ploughing (or any other form of conversion to agricultural land) is prohibited on pain of committing an offence. There are a number of defences: planning consent; or where three months have expired in which no decision has been given; or where twelve months have elapsed following submission of a subsequently rejected application. Thus a planning authority has twelve months in which to consider its response to a proposed extension of agricultural activities. Among the possible responses is the negotiation of a 'management agreement'.[13] Concern in the 1970s at the loss of moorland (aided in part by grants to hill farmers) on Exmoor (North Devon) led to management agreements, negotiated between the Exmoor National Park Committee and local landholders, which were to become the template for more general agreements relating to sites of scientific interest under the 1981 Act.

While a SSSI notice[14] remains in force, any landholder wishing to engage in a listed activity must inform the conservancy council of his intention; he commits an offence if he engages in that activity without informing, or within four months of his so informing, the council. Again, the obligation of 'reciprocal notification' exists to give the conservancy council an opportunity to consider the merits of negotiating a management agreement. This consists of a number of covenants, breach of which gives rise to an action for damages or for injunction to restrain that breach. An agreement is voluntarily entered into, although the threat of compulsory purchase[15] is a tacitly understood (but rarely used) incentive to enter and abide by them. The payment of compensation is a more obvious one; and the Department of the Environment has published advice[16] on the general rules on payments[17] available under the 1981 Act. Lump sums to owners are intended to represent the loss in market value occasioned by the SSSI restrictions, whilst annual sums paid to tenants reflect the reduction in the land's profitability. An agreement will usually provide for the repayment, in the event of a breach, of a proportion of any lump sum received.

Section 29 of the 1981 Act enables a higher level of protection (which may be required by international obligations) to be afforded to particularly sensitive sites by means of a 'nature conservation order' made by the Secretary of State. In the forty or so 'super SSSIs', the period in which the conservancy council may seek to negotiate a management agreement is extended to a maximum of twelve months, and the prohibition on specified operations now applies to any person (not necessarily the landholder). This was the principal point at issue in *Southern Water Authority v. Nature Conservancy Council*.[18] Hill Heath Ditch in the Isle of Wight was designated as a standard[19] SSSI in 1982; five years later a farmer who owned land on one side of the ditch asked Southern Water (who happened to own another area of this SSSI) to dredge the ditch in order to reduce the risk of flooding. Neither party informed the conservancy council and a month passed before the illegal operations ceased. For reasons which did not emerge in the subsequent proceedings, the council chose to prosecute the water authority rather than the farmer. But as the House of Lords ruling made clear, s.28(7) of the 1981 Act applied only to operations carried out by owners or occupiers, and unlike operations on SSSIs designated under s.29, not to contractors. The operation at Hill Heath Ditch, which Lord Mustill was minded to label 'ecological vandalism', therefore went unpunished.

Commenting on the overall effectiveness of the protection offered by the SSSI designation, Lord Mustill averred that 'it needed only a moment to see that the regime was toothless, for it demanded no more from the owner or occupier of an SSSI than a little patience'.[20] The patience he referred to is that required in waiting for the four-month (or twelve in the case of s.29 sites) period to elapse without a management agreement being signed, before commencing the ploughing, tree-felling, drainage or other operation on the designated site.

In view of the 'toothless' reputation which the 1981 Act subsequently acquired, it is difficult now to imagine the unrelenting hostility which it faced in all stages of its passage through Parliament. The book by Lowe[21] details the efforts of the

National Farmers Union and the Country Landowners Association in ensuring that their interests were continually represented in the several hundred hours of debate, and the record-breaking 2,300 amendments, in both the Lords and Commons. Some indication of the febrile nature of the lobbying may be gauged from the fact that 'reciprocal notification' (see above) was presented as evidence of the government's abandonment of what had become known as the 'voluntary principle'. This 'success' for the environmental lobby, led by the Council for the Protection of Rural England, had to be offset by a number of concessions, including the three-month period for landowners to lodge objections against designation. Landowners were to use this appeal period to plough, drain or otherwise destroy an SSSI immediately it was proposed. Repeal of this disastrous provision was urgently needed; it was in fact the primary purpose of a private member's bill which, with government support, became the Wildlife and Countryside (Amendment) Act 1985.

Town and country planning and nature conservation

It must be emphasised that all the categories of protected area considered above are subject to the provisions of town and country planning law (see Chapter 3). The Lake District Special Planning Board exercises both district and county planning functions in its area (which falls entirely within the County of Cumbria); in the Peak District National Park, which spans Cheshire, Derbyshire and Staffordshire, these functions are exercised by a joint board comprised of representatives from these county councils (the majority), from the relevant district councils and some appointed by the Secretary of State following consultation with the relevant conservancy council. In each of the other eight national parks, planning remains the responsibility of the local authorities, but administration of the park is the responsibility of a committee of the relevant county council (or a joint committee where the park spans two or more counties).

Since structure plans were first introduced in 1968, the inclusion in the written statement of 'general proposals in respect of the development and other use of land . . . including measures for the improvement of the physical environment'[22] has been mandatory. The development plans relating to any national park must clearly reflect their special status and the preservation of amenity and recreational opportunities tends to be a dominant, but not always decisive, consideration in the determination of planning applications. In a similar manner, Regulation 37 of the Habitats Regulations 1994[23] (in compliance with Article 10 of the Habitats Directive[24]) requires the inclusion (within structure plans, local plans and unitary development plans) of policies which encourage the management of features of the landscape which are of importance for wild flora and fauna.

In order to preserve the traditional appearance of National Park villages from such intrusions as satellite dishes, plastic cladding and rooftiles of inappropriate colour or material, successive General Development Orders have denied owners of buildings in national parks and areas of outstanding natural beauty the permitted development rights attached to similar buildings elsewhere. Circular 22/80, issued in the first flush of the Thatcherite drive for efficiency in planning (as in other governmental functions), reminds local planning authorities that planning consent should be withheld only 'when this serves a clear planning purpose and economic effects have been taken into account'; however, it goes on to indicate that no change is envisaged in 'the policies in the national parks, areas of outstanding natural beauty and conservation areas'.[25] A recent study casts doubt on the ability of the planning system to protect the parks from development.[26] The minerals which give rise to the rugged geomorphology which attracts climbers and fell-walkers tend also to have a more readily calculable value as raw materials in chemical manufacture. Quarrying – of limestone as well as rarer minerals like fluorspar – in the Peak District would appear to be a perennial source of controversy.

The proximity of a national park should be taken into account when deciding whether a planning application for a proposal falling within Schedule 2 of the 1988 Regulations[27] needs to be accompanied by an assessment of its environmental impact. Circular 15/88 advises that the prospect of significant effects on the special character of an SSSI is another justification for requiring an assessment of a Schedule 2 development.[28] Notwithstanding this ministerial advice, planning authorities enjoy a considerable measure of discretion in weighing the needs of conservation along with other material considerations. The effect of the implementation of the Habitats Directive is considered below, but the ruling in *Beebee* (see Box 8.1) reveals the depth of the English courts' commitment to the discretionary principle in the face of very plausible challenges.

BOX 8.1 R. v. Poole Borough Council ex parte Beebee[29]

Canford Heath is one fragment of that part of eastern Dorset (Thomas Hardy's 'Egdon Heath') which was once a continuous area of bog and heathland; it remains one of the very few locations where the rarest of UK native reptiles, the smooth snake, can be found. Poole Borough Council awarded itself planning consent for housing on part of the Heath six months after the Nature Conservancy Council (NCC) informed them that the land in question was to be brought within the boundaries of an adjacent SSSI. The NCC was subsequently consulted but its objections to the proposal did not sway the planning authority. The legal challenge came not from the conservancy council but from a member of the British Herpetological Society, which had been involved in the site for many years.

Poole BC argued that this body, having no interest in local land, had no standing to sue. Although sympathetic to this view, Schiemann, J. was more influenced by the fact that a condition had been attached to the planning consent which allowed the Society to catch and relocate rare species, and it

---

therefore passed the 'sufficient interest' test which applies in judicial review (as distinct from statutory review under s.288 of the 1990 Act where only persons 'aggrieved' may initiate proceedings). Although granted standing, the Society's representative (Mr Beebee) was unable to persuade Schiemann, J. of the merits of its case.

When approving an application for planning consent, a planning authority is not obliged to justify its reasons (as it is when refusing consent). Third parties are therefore at a disadvantage in challenging an approved application. In this case, the authority had failed to take account of the site's location within a designated SSSI. It was held that this oversight was not decisive since the authority had known of the NCC's long-standing interest in the site. The second grounds of the challenge was similarly dismissed. Under the Assessment of Environmental Effects Regulations,[30] a Schedule 2 proposal with 'significant effects on the environment by virtue of factors such as . . . location' must not be determined without environmental assessment. The proposed housing fell within the terms of a Schedule 2 category (urban development) and its location in an SSSI was clearly 'significant'. But once again, the fact that the planning authority had not considered these points was not sufficient grounds for quashing the planning decision: the planning authority was already aware of the environmental importance of the site and was unlikely to have been diverted from its chosen course by any new information that an environmental assessment might have unearthed.

Alder suggests that the approach adopted in *Beebee* is inconsistent with the 'principle of European Law that domestic law cannot be relied upon contrary to a directive':[31] Article 6.2 of the Directive on Environment Assessment requires that the public must be given an opportunity to express their opinion. In the absence of an environmental assessment, third parties (such as the British Herpetological Society) are denied an opportunity to challenge scientific evidence presented in the decision-making process. That, Alder argues, 'ignores the crucial participatory aspects of environmental assessment'[32].

In commenting upon a similar case,[33] Ward[34] points to the absence of any reference to 'indirect effect' (see Chapter 2) in a ruling in (what was to prove only the first stage of) the Lappel Bank controversy. Swale Borough Council granted Medway Ports Authority planning permission to reclaim Lappel Bank, an area of inter-tidal mudflats in the Medway Estuary. The Royal Society for the Protection of Birds sought a judicial review of this decision claiming that it was in breach of the Environmental Assessment Directive and the EC Directive on the conservation of wild birds. The applicants argued that, if paragraph 1(f) of the regulations[35] implementing the former, which refer to 'the reclamation of land from the sea . . . for the purposes of agriculture', were interpreted so as to include land reclaimed for other purposes, it would be more consistent with the aims of the original directive and would then oblige the planning authority to consider an environmental assessment. Simon Brown, J. was aware that other types of land use cited in the Directive had been omitted from the regulations. He did not consider that it fell to him to redress that omission. He abided by a literal interpretation of the regulations, since accepting that proposed by the applicants required 'torturing [their] construction beyond breaking point'. For this judge, the question of whether of an assessment was needed was not a matter of law but one of fact and one which was 'exclusively for the planning authority subject only to *Wednesbury* challenge'.[36]

The Habitats Regulations

The objectives of the Habitats Directive[37] (92/43/EEC) will be met primarily by the designation of a number of 'special areas of conservation' (SAC) within which specified flora and fauna will be protected. Each member state of the European Community will contribute, in proportion to the occurrence of listed species and habitats within its territory, to a Europe-wide network of SACs, known as 'Natura 2000'. The great majority of the British contingent of SACs[38] (the term will also embrace the 'special protection areas' (SPAs) required by the 1979 Wild Birds Directive) will also be SSSIs under the 1981 Act.

The protective aim of the Directive extends to requiring member states to review licences, authorisations and consents for activities which could lead to the deterioration of a European site. The regulations[39] which implement the directive in the UK refer to a number of statutory authorisations relating to highways and pipelines, to pollution authorisations and to waste management licences (Parts I and II of the Environmental Protection Act 1990). Compliance with the Directive in Great Britain has primarily involved amending the law covering the control of development under the town and country planning system.

The 'Habitats Regulations' restrict the discretion of planning authorities when considering applications for planning consent for development which, following a scientific assessment, is adjudged to 'adversely affect the integrity' of an SAC. The planning authority must first satisfy itself that no alternative solution exists. If no alternative is identified and the site does 'not host a priority natural habitat type or species', then permission may be granted only for 'reasons of overriding public interest, including those of a social or economic nature'. Regulations 60–3 ensure that proposals likely to have a significant effect on a European site and which would normally constitute 'permitted development'[40] are not begun without the approval of the local planning authority. In addition, planning authorities are required to conduct a review of extant planning permissions which could have a significant effect on an SAC and, if necessary, revoke, modify or discontinue such permissions (notwithstanding the possible obligation to pay compensation).

Priority natural habitat types and species are listed under Annex I and II of the Habitats Directive; they include, in the UK, limestone pavements (which were the subject of specific protection under s.34 of the 1981 Act) and the Caledonian forests. Where development is proposed for which no alternative is apparent and which would adversely affect the integrity of such a host site, then the only considerations which can justify the grant of planning permission are:

- human health or safety
- beneficial consequences of primary importance to the environment; or
- other imperative reasons of overriding public interest, following consultation between HM Government and the European Commission.

The 'overriding public interest' proviso may be a reaction to the European Court of Justice's strict interpretation of Article 4.4 of the Birds Directive in *Leybucht Dykes*.[41]

The dykes in question have for centuries protected the low-lying East Frisian coastline. A wetland of international importance for both listed and migratory birds, it was designated an SPA in 1985. In carrying out civil engineering work to strengthen the sea defences, the Lower Saxony *Land* disturbed some birds and reduced the area of their habitat. This prompted the Commission to initiate an Article 169 action against Germany in the European Court. Here Germany, with the support of the United Kingdom, argued that the duty 'to take appropriate steps to avoid pollution or deterioration of habitats affecting the birds' within a designated SPA was qualified by an obligation to take 'account of economic or recreational interests' (the phrase used in Article 2 of 79/409/EEC to qualify a member state's general duty to conserve all wild bird species). This was rejected by the Court, but it conceded that derogations of Article 4.4 – the duty to prevent pollution, deterioration and disturbance of designated habitats – might be permitted in exceptional circumstances if they entailed 'a general interest superior to the ecological objective envisaged by the Directive'.[42] The reduction of risks to human life, by maintaining sea defences, was held by the Court to satisfy the 'general interest superior' criterion in this particular instance.

Thus a clear judicial interpretation of Article 4.4 outlaws disturbance of a designated SPA in all but exceptional circumstances. One consequence of this ruling may be to shift attention to the pre-designation stage. A wish to retain the option of subsequent development in or near a site may result in fewer or smaller sites being put forward, or their boundaries being drawn by matters other than purely ornithological. This latter point was central to the RSPB's judicial review of the Secretary of State's decision over Lappel Bank.

Lappel Bank consists of twenty-two hectares of partially reclaimed land from a proposed SPA of 4,681 hectares in the estuary of the River Medway in Kent. During extensive consultations, the Port of Sheerness Limited stressed the importance of the Bank and adjoining land in the proposed expansion of the port, for which planning permission had been granted (see above). Although aware of the ornithological value of the site, the Secretary of State concluded that it was outweighed by the economic opportunities, and he drew the boundaries of the designated site with Lappel Bank excluded. (Cardiff Bay, a bigger but similarly contested site in Wales, was excluded totally from consideration.) In the High Court (and on appeal to the Court of Appeal and the House of Lords), RSPB argued that Lappel Bank provided good quality feeding grounds for a number of waders and wildfowl and was a small but important part of the total ecosystem. On the proper construction of Directive 79/409 and in the light of recent European case law, notably *Leybucht Dykes* but also the *Santona Marshes*[43] case, only ornithological considerations were pertinent at the designation stage. This contention was rejected by the Divisional Court and by the Court of Appeal (but with Hoffman, L. J. dissenting); the House[44] resolved to refer the matter to the ECJ under Article 177.

In July 1996 the ECJ[45] accepted the Advocate General's earlier opinion,[46] which followed the *Leybucht* precedent in arguing that a member state is not entitled to take economic requirements into account when classifying SPAs or determining their boundaries. The Court avoided a further elaboration of the 'general interest which is superior to . . . the ecological objective of the Directive'[47] but insisted that economic requirements could not qualify. In addition, 'economic requirements [amounting to] imperative reasons overriding public interest'[48] must be excluded from consideration at the designation stage, even though they might be cited later to justify encroachment.

The Lappel Bank decision is clearly very important and could have implications for the designation of other habitats, not simply those of birds. Non-human interests are now firmly established as the primary (if not absolutely exclusive) considerations both ante- and post-designation.

The new regime of habitat protection can now make inroads into the rights (immunities) which 'permitted development' status has hitherto conferred upon agriculture. It was the thinly veiled threat of a wider use of planning powers to remove permitted development which forced the coalition of farmers and country landowners to accept the principal provisions (the 'voluntary principle') of the 1981 Act. The Department of the Environment's threatened use of a planning direction, to prevent ploughing of wetlands on Halvergate Marshes, eventually persuaded the farmer at the centre of this *cause célèbre* to accept a management agreement in 1984.[49] But the protection now afforded to SACs by the Habitats Regulations is of a different kind: the planning authority is not required to issue a direction because permitted development is automatically suspended[50] in the case of any agricultural operations which might have a significant effect upon a Natura 2000 site.

The Habitats Regulations comprise a regime which seems unlikely to share the 'toothless' reputation of standard SSSIs. One word of caution is necessary. It is possible to point to seemingly innocuous actions by farmers which, because they do not amount to 'operations' which are specified as permitted development, cannot be proscribed by the new regulations. For instance, simply opening a gate and allowing sheep to move from one field to another is not an operation (like ploughing or uprooting a hedgerow) which can be outlawed by suspending permitted development rights; but that action could prove equally devastating for a colony of rare plants. The new regime has not entirely replaced the 'voluntary principle'; for its success still depends upon farmers giving it more than their grudging cooperation.[51]

A statement that voluntary management agreements and monetary incentives are insufficient to secure compliance with the Wild Birds Directive also formed part of the *Leybucht* ruling in 1991.[52] If, as I am suggesting, this case demonstrates the onerous nature of member states' obligations over designated habitats, it is not surprising that the relevant Planning Policy Guidance[53] (published three years after the ECJ ruling) indicates that the Secretary of State will normally 'call in' and determine himself applications which are likely to significantly affect sites of international importance. Revocation of an existing planning consent (for any class of development) is a rare occurrence; it is therefore necessary to note that in March 1991 the Secretary of State (Michael Heseltine) announced his decision[54] to revoke Poole Borough Council's planning consent for housing on Canford Heath SSSI. Dorset Heathland (of which Canford Heath is a part) is now listed among potential sites for SAC status; in fact, the RSPB has suggested that, as an example of a Southern Atlantic Wet Heath, it should be treated as a 'priority habitat'.[55] It is also one of eight SSSIs within English Nature's 'Wildlife Enhancement Scheme' which meets the costs of fencing, cattle grids and other measures necessary to prevent further scrub invasion of the 5,000 hectares which is all that remains of this extremely rare habitat.[56]

Since the Habitats Regulations were introduced under the authority of s.2(2) of the European Communities Act 1972, they required a resolution of both Houses of Parliament before coming into effect. It would not be correct to say that the ninety-minute Commons debate on the draft regulations and the two-hour debate in the Lords were mere formalities. In both cases, discussion tended to be dominated by individual concerns such as subsidiarity and apprehension over the 'Henry VIII clauses' (see Box 8.2) contained in the regulations. Front-bench speakers were repeatedly required to give assurances that voluntarism remained the guiding principle of HM Government's policy on conservation and rural land use. But the readiness with which these assurances were given does not alone account for the absence of a more determined opposition to these proposals. Although the major protagonists of the epic struggles which accompanied the passage of the 1981 Act were again represented, no desire to resume hostilities in earnest was apparent. The marked 'greening' of the Community's Common Agricultural Policy in the intervening period may form part of the explanation.

Agriculture

The need for the integration of an environmental dimension in all Community policies was accepted in the Third Action Programme on the Environment in 1983. If Council Regulation 797/85,[57] on improving the efficiency of agricultural structures, can now be cited as an early exemplar, that was only because of

---

BOX 8.2 'Henry VIII clauses'

The Statute of Proclamations 1539 empowered the sovereign to legislate by decree; the term 'Henry VIII clause' is now applied in a derogatory sense to any statutory instrument which empowers a minister to amend primary legislation. The Habitats Regulations amend neither statute nor statutory instrument. In the Lords, the term was used by Lord Williams of Elvel because he felt that, with its seventy-five pages and its provisions for the Secretary of State to make special conservation orders and for the conservancy councils to make byelaws, the Regulations should have been a Bill (House of Lords Official Report [Session 1993–4] vol. 558, 17 October 1994, cols. 85–116). In the Commons, the member for Cardiff West used the term when deprecating the absence of guidelines to assist the determination of 'overriding public interest' by the Secretary of State and, in regard to priority sites, the European Commission (House of Commons Official Report [Session 1993–4] vol. 247, 19 July 1994, cols. 239–61). Notwithstanding the fact that convention dictates that neither House may amend a statutory instrument, to apply the term 'Henry VIII clause' to any provision in a debate in which it (and the rest of the instrument) could be annulled seems, at very least, an historical solecism.

---

pressure from a UK source.[58] When considering the draft regulation in 1984, the House of Lords Select Committee on the European Communities had the benefit of evidence from the Council for the Protection of Rural England, the Institute for European Environmental Policy and the Countryside Commission, in addition to the National Farmers Union and other bodies representing the food industry. That the NFU did not prevail on this occasion may be inferred from the fact that the Select Committee found the draft regulation to be 'backward looking' and too concentrated upon production factors, with too few measures directed towards the diversification of rural activity and enhancement of the environment. The Committee's recommended amendments were broadly accepted by the Minister of Agriculture who, despite opposition from other members of the Council of Ministers, was eventually able to ensure that Article 19 of the Regulation made a significant step forward in encouraging farming practices which conserved habitats or enhanced the landscape.

Implementation of Article 19 required the passing of the Agriculture Act 1986, s.18 of which empowers the Minister of Agriculture to designate, by statutory instrument, an area as an 'environmentally sensitive area' if that appears desirable to:

- conserve and enhance the natural beauty of the area; or
- conserve the flora or fauna or geological or physiographical features of the area; or

- protect buildings or other objects of archaeological, architectural or historical interest in the area.

The first five areas to be designated in England, and their respective rationales, are listed in Box 8.3.

BOX 8.3 Environmentally Sensitive Areas: the first five

- Norfolk Broads – to prevent further drainage and conversion to arable;
- Pennine Dales – protecting traditional stone-walled hay meadows from reseeding and silage cultivation;
- Somerset levels – as for Norfolk Broads;
- Eastern End of Southern Downs – to conserve grazing (and therefore the characteristic flora) on chalk downland increasingly depleted by conversion to arable;
- West Penwith (Cornwall) – to protect its prehistoric settlements and its ancient field patterns from intensified dairy farming.

The Ministry of Agriculture, Fisheries and Food (MAFF) enters into agreements by which owners and occupiers of land (and any successors in title) within an ESA receive financial 'incentives' to desist from damaging practices or to revert to more benign forms of cultivation or husbandry. Unlike the SSSIs discussed earlier, it is MAFF and not the Department of the Environment which enters into ESA agreements in England. Moreover, the terms of proposed management agreement for an ESA are laid down in the relevant designation order, since they are dependent upon the particular area and its characteristic form of agriculture. The Minister is not compelled to offer an agreement, and in this sense the ESA scheme is more 'voluntary' in character.

The ESA provisions are funded from national sources, not the Community agricultural budget, and hence the need for approval by HM Treasury before an area is designated. However, Community funding has been available since 1988 to compensate farmers who take land out of arable production; £11 million was allocated for payment to UK participants in 1988–9, the first year of the 'set aside' scheme.[59] Although it was primarily a measure to curb the Community's embarrassingly large 'grain mountain', set aside has, like the earlier 'milk quotas' which also discourage intensification of agricultural land use, an implicit environmental dimension. Set-aside land can be given over to nature reserves and woodland. The use of fertilisers is prohibited on land set aside as fallow, and the use of pesticides requires special permission. Thus the fallow option might encourage the movement to organic farming.

The Common Agricultural Policy remains by far the largest single item of Community expenditure, and it has defied all attempts at wholesale reform. The early success of 'set aside' has encouraged further measures which couple environmental with supply-reducing objectives. Again, the general feeling of the House of Lords Select Committee on the European Communities, when considering the latest EC Regulation, was supportive of further financial inducements to promote 'extensification' in agriculture. Evidence presented by the Council for the Protection of Rural England raised a note of caution:

> There is increasing acceptance that farmers should be paid for their role in *providing* landscape management and other environmental services; and that financial incentives can play an important role in encouraging farmers to *convert* to more environmentally positive farming practices; but these should not be confused with payments to *desist* from environmentally damaging activities. Such payments conflict with [the polluter pays principle], particularly if the aid is more than transitional. Compatibility with the PPP should be one of the criteria on which schemes proposed for Community aid are judged.[60]

In Chapter 5 passing reference was made to the 'nitrate sensitive areas' (NSAs) designated in the aftermath of the ECJ action on the UK's non-compliance with the Drinking Water Directive (80/778/EEC). Further consideration is given here because although the aim of NSAs is undoubtedly directed to the protection of resources of potable water, they only affect agricultural activities.

In 1990 ten pilot NSAs were introduced[61] by way of an experiment to gauge the effectiveness of measures designed to reduce concentrations of nitrate in ground- and surface waters. In July 1994 a further twenty-two NSAs were instituted[62] by way of implementation of EC Agri-Environment Regulation 2078/92. The options available differed between arable and grasslands, but they shared the underlying aim of nitrate reduction. This aim was pursued directly by limiting the application of artificial fertiliser and organic manure (for which codes of good practice have been drawn up) and indirectly by encouraging a less intensive use of grassland and the sowing of varieties of grass which 'lock in' nitrate. Since July 1995 it has been possible for arable land taken out of production under the NSA scheme to count towards a set-aside obligation. It is recognised that adhering to these various measures results in reduced yields; compensation is payable in respect of income foregone to farmers who enter into NSA agreements (now replaced by 'undertakings' which do not entail a legal charge being registered on the land) with MAFF.

It seems unlikely that the UK government's 'voluntary' approach will form an adequate basis for compliance with the 'Nitrate Directive'.[63] This requires member states to designate as 'nitrate-vulnerable zones' (NVZs) all known areas of land which drain into surface or groundwaters where the 50mg per litre limit for nitrate in drinking water is (or is likely to be) exceeded. Marine and freshwater bodies found to be eutrophic (rapid plant growth resulting from an excess of phosphate and nitrate nutrients) may also be designated. All thirty-two existing nitrate-sensitive areas fall within the sixty-eight provisionally designated NVZs (covering some 8,000 farms and about 600,000 hectares in England and Wales). Participation in NSA undertakings is voluntary, but the various control measures applied to an NVZ are compulsory. The nitrate-vulnerable zones will be instituted by order[64] which will set out 'requirements, prohibitions or restrictions' concerning agricultural practices. The order may also provide for payments to be made in respect of the performance of certain associated obligations. Although grants may be available towards the purchase of appropriate waste storage and other facilities, the earlier principle of compensation for lost profit appears no longer to apply.

Discussion

The need to distinguish a category of environmental rights from those which follow from the ownership of property has been stressed in earlier chapters. This task is not assisted by the number of instances, cited in this chapter, in which the state reveals a deep-seated aversion to interference in those rights which stem from an interest in agricultural land or forestry (see Box 8.4). This aversion tends to be coupled with a commitment to compensation for any such interference which is permitted. In the UK, as in other member states, a large proportion of that compensation originates from Brussels. The European Community has had a Common Agricultural Policy (article 39 of the EEC Treaty) since its foundation in 1956; but despite repeated calls, the CAP has defied most attempts at reform. It remains to be seen whether the CAP's recently acquired 'green' dimension can lead to the fundamental change in priorities needed to make inroads to its massive budget. However, it seems unlikely that the CAP, as currently conceived, can survive the accession of Poland, Hungary, Estonia, Slovenia and the Czech Republic to the European Union.

As the Council for the Protection of Rural England pointed out in their evidence to the House of Lords,[65] there is a difference between payments for measures which improve the environment and compensation for abstaining from practices, like nitrate use, which are harmful. In theory, the latter violates the 'polluter pays' principle and is, following the amendments introduced by the Single European Act in 1987, contrary to the Treaty of Rome. In view of the failure (see Chapter 2) of the attempt to rely upon the indirect effect of a comparable element of Article 130r (namely the 'precautionary principle') it would seem that legal challenge of such payments in a national court is likely to founder on similar grounds. Whether the law should view an activity – like the spreading of fertiliser – as an act of pollution to be penalised or as a traditional farming practice, interference with which merits compensation, is ultimately a question for the legislature. The fact that the UK parliament can contemplate the former is testimony to the eclipse (albeit partial) of the farming lobby.

It should be remembered that the encouragement of artificial fertilisers was partly a response to wartime memories, in the UK and elsewhere in Europe, of the strategic benefits of improved productivity and reduced reliance upon food imports. Recognition of the environmental disbenefits of this policy are far more recent. The

173

immunities which agricultural activities have enjoyed, compared with industrial polluters of watercourses, cannot be explained simply by the fact that nitrates and slurry are inherently less toxic and persistent than oil or organohalogens. The 'industrialisation' of agriculture – of which BSE is perhaps one of the most distressing symptoms – may account for a greater political readiness to erode, albeit slightly, those immunities. This may be part of a wider acceptance that the rights which come with ownership of land are subordinate to a deeper public interest.

The protection of potable water supplies is undoubtedly a public interest; but it remains an anthropocentric one. The Habitats Directive necessitates a deeper intrusion on traditional property rights and one which has a distinct ecocentric motivation.

The Habitats Directive (as may any other promulgated after Maastricht) refers directly to the objective of 'maintenance of biodiversity' and justifies that reference by its contribution to the 'general objective of sustainable development'. The hierarchy – in which habitat preservation is situated below human health, and safety above economic and social concerns which fail the 'overriding public interest' test – is not to be taken lightly, especially in the light of the European Court's rulings in the *Lappel Bank*[66] and *Leybucht Dykes*.[67] If the Community's institutions come to adhere to one of the variants of 'strong sustainability' and insist upon a vigorous defence of biodiversity and similar forms of 'natural capital',[68] then ecocentric concerns could become further embedded within European Community law.

To committed advocates of animal rights, these points will be dismissed as trivial and serving only to divert attention from the iniquities of a European Community which has yet to ban, among other atrocities, the live export of veal calves. Such arcane arguments, they might continue, serve only to emphasise the need for civil disobedience and extra-judicial means in the struggle to protect sentient creatures. It is tempting but misleading to see the civil rights which we currently enjoy in the UK as the product of isolated and momentous events: Magna Carta, the Bill of Rights of 1689, the 1832 Reform Bill. In the absence of a written constitution, the development of 'rights' recognised in English law has been characterised by incrementalism. According to one theorist, a human right 'is a conceptual device . . . that assigns priority to certain human or social attributes'.[69] If there are circumstances in which *priority* is assigned to the protection of non-human species rather than the pursuit of human interests, then even if that assignment does not amount to the recognition of a clear ecocentric right, it does represent some erosion of the anthropocentric hegemony. The Habitats Regulations represent a small step forward but they are not insignificant, even if other jurisdictions (see Box 8.4) can claim to have made greater progress. They clearly state the *priority* to be attached to the protection of the most important habitats – below that of human health and public safety but above that of economic efficiency. Habitat protection has therefore acquired that 'trump' quality which Dworkin[70] argues is characteristic of rights. Of course, it can be over-trumped; but that is another quality which it shares with human rights.

BOX 8.4 **The spotted owl *v.* the timber industry**

Under the United States Endangered Species Act of 1973, the Secretary of State for the Interior, acting on the advice of the Fish and Wildlife Service, can 'list' species of animals and plants in danger of extinction. It is then an offence to 'take' a member of a listed species. The term 'take' is defined (by 16 USC §1532 (19)) to mean 'harass, harm, pursue, hunt, shoot, wound, kill, trap, capture, or collect, or to attempt to engage in any such conduct'. A regulation made by the Interior Department in 1975 to implement the statute further defines 'harm' to include, in addition to acts which kill or injure, 'significant habitat modification or degradation where it actually kills or injures wildlife by significantly impairing essential behavioral patterns, including breeding, feeding or sheltering' 50 CFR §17.3 (1994).

The plaintiffs in a challenge to this definition of harm, which was eventually decided by the US Supreme Court, were all connected in some way to the timber industry. Since the dwindling numbers of the northern spotted owl are concentrated in the forests of the north-western US, this industry is particularly sensitive to a strict enforcement of the ban on modification or degradation of the habitat under the 1975 regulation. By a majority of 6–3, the Court dismissed the challenge, leaving owners of private property in no doubt that activities, on which their livelihood may depend, become unlawful when they affect the habitat of any one of (now) a thousand endangered species. The 1973 Act contains no provision for compensation, nor does it contain any defence comparable with the 'overriding public interest' which appears in the closest European Community analogue, the Habitats Directive. And since the US Act 'lists' more species and is not confined to 'special areas of conservation' (in fact it extends to US territorial waters), it must represent a far deeper commitment to species protection than its European counterpart.

But the good intentions of Congress in 1973 and the administration in 1975 may count for little, if and when this Act, with its uncompromising defence of habitats, is reviewed by a Congress no more immune to the 'environmental backlash' of recent years than those who elected it. In which case, the less dogmatic approach of the European Community – necessary if a qualified majority is to be secured despite the diverse views of fifteen member states – may prove to be of more lasting benefit.

Source: Bruce Babbitt, Secretary of the Interior, et al., petitioners v. Sweet Home
Chapter of Communities for a Great Oregon et al. (United States Supreme Court, 29 June 1995, 115 S Ct 2407, L Ed 2d 597 1995); and the analysis of this case by Michael Herz (1996) 8 *JEL* 179.

9

ARE ECOCENTRIC RIGHTS SO VERY DIFFERENT?

Should we protect animals?

Had Disney chosen a less popular site than Mineral King Valley, then *Sierra Club v. Morton*[1] might never have occurred; and environmental philosophy would have developed without the insights of Christopher Stone's celebrated essay 'Should trees have standing?'.[2] Had no comparable conflict arisen earlier, then *Lappel Bank* might have stimulated a European environmentalist to ask the question: 'Do ducks have rights?'. And since this later challenge by Europe's largest wildlife charity was ultimately successful,[3] whereas the Sierra Club's fell by the narrowest (3–4) of margins in the US Supreme Court, the obvious temptation would be to answer 'yes'.

But I believe the answer to be less important than the particular circumstances which gave rise to the question. Philosophers may ponder whether animals, trees or ecosystems can ever be the holders of rights, but the courts are called upon to consider less abstract questions: first whether an individual or organisation, dedicated to some non-human constituent of the natural world, can be allowed to make his case; and second, whether that case is more persuasive than any of its rivals. The Royal Society for the Protection of Birds was allowed to make its challenge and was successful:[4] the Secretary of State had acted unlawfully in drawing the boundary of the 'special protection area' (SPA) to take account of economic considerations. The RSPB was given the ruling it sought, but throughout the proceedings the suggestion that it was necessary to protect the 'rights' of the waders and other wildfowl never arose. However, the influence of the rights discourse is such that many would summarise this case as one in which the 'rights of birds' took precedence over those of the developers.

The RSPB's role as *de facto* 'guardian' of the UK bird population was undoubtedly strengthened by this case. *Lappel Bank*, and any subsequent cases in which it is taken as authoritative, serves to reduce the risk of future encroachment upon any designated SPA. It is necessary to recall that this concern with habitats comes after more direct protection via statutes which make it an offence to kill, injure, take any wild bird, its eggs or its nest (while in use). Where necessary, the RSPB has taken private prosecutions in the criminal courts to penalise egg collectors and deter others.

Another organisation, the League Against Cruel Sports, has sought to protect red deer using one of the ancient property rights which is more often employed *against* environmental activists.[5] As the owner of land on Exmoor, the League was able to take action[6] in trespass against the Joint Masters of the Devon and Somerset Staghounds, following incursions onto its land by hounds, which the masters had been either unwilling or unable through negligence to prevent. Although a local authority is not statutorily disabled from similarly setting up 'sanctuaries' on land in its possession, it must do so for the purposes of 'benefits, improvements or development'[7] of those areas, and not simply on account of its members' moral repugnance at blood sports (see Box 9.1). No such restriction applies to the National Trust. With 240,000 hectares, it is one of the largest landowners in England and Wales; its decision in March 1997 to ban stag hunting on its land followed the publication of a report by an eminent zoologist which concluded that hunting with hounds caused deer to experience great and unnecessary suffering. In a subsequent judicial review of this decision brought by the Master of the Quantock Staghounds, the judge refused to lift the ban but deprecated the 'speed and secrecy' with which the decision had been taken and urged the Trust to reconsider it.[8]

BOX 9.1 R. v Somerset County Council, ex parte Fewings

In 1974 Somerset County Council, under s.122 of the Local Government Act 1972, acquired some land at Over Stowey Customs Commons, over which red deer were regularly hunted by hounds. Section 120 of this 1972 Act provides: '(1) For the purposes of . . . (b) the benefit, improvement or development of their area, a principal council may acquire by agreement any land whether situated inside or outside their area'. After a heated debate, those councillors opposed to hunting on moral grounds secured a 26–22 majority to ban hunting on the Commons. A representative of the Quantock Hunt sought to challenge this decision by judicial review.

For Mr Justice Laws,[9] the central question was whether the power delegated to a local authority to acquire (and manage) land was wide enough to allow moral perceptions to fall within the notion of 'the benefit, improvement or development of their area'. The interference with the lawful liberty to hunt stags could be justified only if the Council were satisfied of an objectively demonstrable link between that prohibition and the 'benefits . . . of their area'. That hunting with hounds was seen by the majority as morally repugnant was incidental to that link; therefore the ban was unlawful. In the subsequent hearing in the Court of Appeal, the Master of the Rolls could not accept Laws, J.'s earlier contention that the cruelty argument was *necessarily* irrelevant to a calculation of the benefit to the area. That would

---

177

require too narrow an interpretation of the wording of s.120 of the 1972 Act. In other words, it was possible to imagine a situation in which a moral consideration could materially influence the link. An examination of the circumstances of this case (especially the minutes of the County Council debate) failed to show that the councillors had been advised of the need to identify such a link.

The application was dismissed by a 2–1 majority. The dissenting views of Simon Brown, L. J. are of particular interest. He accepted that:

> the concept of benefit to the area, and public interest and good, invite considerations first of the Council's human community rather than its wildlife. But the two considerations are not discrete; human well-being for many will depend upon their satisfaction as to animal welfare.[10]

He agreed with the Master of the Rolls that Laws, J. had been in error in claiming that the cruelty issue was necessarily irrelevant but Simon Brown, L. J. went further and argued that on the contrary, this, along with other ethical arguments presented in the debate, was necessarily relevant to the decision. If the case is heard in the House of Lords (leave to appeal was granted to Somerset County Council) then the minority view will no doubt be further rehearsed. It is necessary at this stage to comment that Lord Justice Brown's argument should not be seen as an affirmation of 'animal rights'.

It is perfectly reasonable to claim that the well-being of a majority of the inhabitants of any local authority area is indeed contingent upon their being satisfied that the welfare of deer (or any other animals in that area) has been safeguarded. Moreover, such a claim could serve to justify the use of a discretionary power to effect appropriate safeguards (for example, the purchase of land for 'deer sanctuaries'). It is the pursuit of *human* utility which is the prime-mover of this process, in which animal welfare is the means, not the end in itself. Therefore it remains firmly in the anthropocentric camp. A rights-based approach would require the abolition of hunting irrespective of whether such a ban added to or subtracted from the sum of human utility.

I am unable to identify a provision in current English law equivalent[11] to that by which the Spanish Supreme Court awarded damages to the Asturian Fund for the Protection of Wild Animals after a bear was shot by a hunter.[12] However, it is not fanciful to predict a greater statutory recognition of wildlife and animal welfare groups, given the recent relaxation of judicial attitudes to the standing of non-governmental organisations (NGOs) in public law actions. Without underestimating the political resistance which the agricultural and other lobbies might mount, it is not difficult to imagine the gradual extension of both statutory and common law remedies enabling state bodies, NGOs and individuals to ensure effective protection of increasing numbers of species and in ever wider areas of the country.

Rather than endlessly debating whether animals can meaningfully be said to be the holders of rights, one might ask: 'within jurisdiction Y, are the interests of species X more effectively protected than those of children?'. We then have a question which (laying aside the practical problems arising from the greater numbers of children) admits the possibility of empirical test. When X stands for ospreys and Y for Scotland, my awareness of the RSPB's determination to make the eyries at Loch Garten (and elsewhere on Speyside) impregnable persuades me to answer in the affirmative. When X stands for badgers and Y for England, then my response would have to await the results of studies of the badger population and of the numbers of prosecutions brought under the Badgers Act 1992. These data would allow me to form an assessment to be compared with the one which I have made, first as a child myself, later as a parent, and in more recent years as one of many of UK inhabitants appalled to witness the succession of scandals concerning sexual and other abuse of children, especially those in the care of the state.

How can we claim to know (far less measure) what the interests of badgers might be? The observations of naturalists would suggest that food, shelter from the elements and safety from predators represent an obvious starting point. That may well not be an exhaustive list; if they also need the equivalents of love, affection, praise, encouragement or whatever, then other badgers are, I suspect, best placed to provide these (whether consciously or not), and all that any other badger-regarding species can do is ensure that the three cited needs are satisfied and then simply respect the privacy of the sett. The more complex the species, the greater the variety of interests and the greater the possible harm that may be inflicted, even when it is members of the same species who decree the interests of their fellows. Claiming a knowledge of the interests of each and every member of society is the hallmark of totalitarian regimes of the left and right alike. Regardless of the amount of thought devoted to the question, parents must ultimately have some conception of the interests of their children; acting arbitrarily is incompatible with the exigencies of childcare. Indeed, parents' responsibility to care for their children now has legal[13] as well as moral force; and organs of the state assume that duty very reluctantly and only when it is irrefutably demonstrated that parents (whether natural, adoptive or guardians) are unwilling or unable to discharge it themselves.

Over centuries, the Court of Chancery developed rules to safeguard the rights (especially to property) of orphaned children. The fact that English law now dictates that the interests of the child shall be paramount[14] in relevant decisions of the Family Division of the High Court does not make its deliberations any easier. Quite apart from the decision as to who shall care for the child, other agencies of state must ensure that adequate care is duly offered. My point is that the interests of children, like those of animals, can be thought of as 'social facts'[15] which, although they cannot be measured like height, weight or blood pressure, exist in the real world independently of an observer's perceptions. Apart from the paramount position of the child's interest, there are no further guiding principles; pragmatism is all-important in an effort to optimise the well-being of those too young to pursue it for themselves. A bruise, whether on the arm of a two-year old child or on the flanks of a horse, must be taken as *prima facie* evidence of a criminal act unless an innocent explanation is forthcoming. We have created a legal framework which aims to protect the interests of children; and we are in the process of creating a comparable framework for selected animal species. The more similar[16] the two frameworks, the more we are justified in the mutual exchange of their respective vocabularies. I suggest that we may have reached the point where the similarity is such that the deliberate use of separate vocabularies would amount to a distortion of language: whatever the elusive quality that gives children their rights, then certain animal species must be said to have it as well.

For Professor MacCormick, it is uncontestable that 'at least from birth, every child has a right to be nurtured, cared for, and if possible, loved, until such time as he or she is capable of caring for himself or herself'.[17] He declares this to be a moral right and, whilst conceding that 'many legal systems' have not recognised it as a right, goes on to claim that ours (which I take to mean Scotland and England) currently do so. Since I do not intend to take issue with either point, and since I am disinclined to criticise his use of the word 'right', it must follow from my 'similarity' argument above that I could not contradict a statement that the osprey chicks at Loch Garten enjoy a similar right to be nurtured and cared for. Golden eagles, roseate terns and Scottish crossbills are other bird species whose rarity necessitates protective measures which give them a similar status. Among mammals, all fifteen species of bat occurring in Britain enjoy a level of protection which can also be said to amount to a 'right of undisturbed existence'. As we consider more and more species, there will clearly come a point where the degree of protection or the rigour of its enforcement is such that the attribution of that right is no longer justifiable. But is it possible to identify general dissimilarities, common to all non-human species whether rare or plentiful, which undermine this argument?

The fact that animals are unaware of, and cannot themselves invoke, any rights they might hold, being inevitably reliant upon 'guardians' like RSPB, does not qualify. The same can be said of young children (albeit their parents or guardians are from the same species) and that is precisely why I, following other writers, have chosen children's rights for comparison.

But could there not arise conflicts when several NGOs claimed guardianship over whales, seals, golden eagles, natterjack toads or whatever? I'm sure that disputes could occur but I do not see that they would necessarily prove fatal to the cause of animal protection. It is possible to imagine some body fulfilling a regulatory role comparable with that of the Charity Commission. The Charities Act 1993 makes clear that an organisation devoted to animal, rather than human, welfare may still be a 'charity' under the terms of that Act. Moreover, rules over contested (human) wardship which evolved in the Court of Chancery might offer initial guidance.

Is the protection given to certain animals in some sense more revocable than that enjoyed by children? No parliament may bind its successors; any statute can be repealed. Overturning a statute, so deeply entrenched as to acquire a 'constitutional' appearance, will require a far greater determination by the ruling majority if it is to survive the full parliamentary process. Statutes passed to implement international treaties and conventions may, if repealed, entail international opprobrium, if not sanctions; but those concerned with children do not necessarily attract special attention. Attempts by Thatcher governments in the 1980s to resist European Community restrictions on the maximum working hours of minors were couched in the rhetoric either of free enterprise or of interference by Brussels bureaucracy; reference to possible conflict with the UN Convention on the Rights of the Child (which Mrs Thatcher had earlier signed) was rarely made.[18] International treaties on wildlife protection and endangered species certainly outnumber and predate those concerned with the rights of children.

Progress towards the level of protection which certain species now enjoy in the United Kingdom has been gradual and piecemeal; on some occasions international influences have initiated change, on others, reform has been driven by domestic concerns. But the history of animal protection is devoid of momentous struggles analogous to those which preceded the Bill of Rights in 1689, or to cite an American example, emancipation of the slaves in 1863, which have marked the development of (human) civil rights. Moreover, the progress has varied considerably from species to species. The question often raised by sceptics, 'Do spiders have rights?' serves to remind us that the great majority of invertebrate species enjoy no protection at all. However, the fragmentary and incomplete character of animal protection does not disqualify it from a rights discourse. Despite our tendency to assert human rights to be universal, fundamental and inalienable, we do not feel obliged to abandon the language of rights when reminded of societies in which they are regularly violated or not recognised at all. The existence of slavery in Ancient Rome does not prevent historians describing the hierarchy of rights by which other social classes were defined. For centuries, debates about the rights of *man* took place with no thought being given to the need to challenge the legal status of wives as their husbands' chattels. I do not ascribe less value to the civil rights which I currently enjoy because of my awareness that they could be suspended tomorrow (if, for example, the Crown used its prerogative to assume emergency powers).

Although I am unprepared to contradict an assertion that there are animal species in the United Kingdom which have 'a legal right to life', I believe that utterances of this type should be used sparingly. For they tend to divert concern away from the constant need to minimise violations and to punish the perpetrators. I am prepared to accept that some whales can be said to have a right to life under international law. But this observation must not be allowed to detract from the urgent need to reduce quotas on the commercial and 'scientific' killing of some species, to extend protection to others and to eradicate 'pirate whaling', i.e. by vessels flying flags of convenience in order to escape the controls of the International Whaling Convention.[19]

Should we protect plants?

It is an offence to pick or uproot any of the species of rare wild plant.[20] The Habitats Directive is concerned with endangered flora no less than fauna. Lady's-slipper orchid (*cypripedium calceolus*) survives at only one location in northern England; English Nature has taken such extensive measures to ensure its survival that attributing a 'right to life' to this colony (or to the three sites in Norfolk to which the Fen orchid (*liparis loesilli*) is now restricted) is not to use a metaphor. I see no reason why wildlife NGOs should not be given recognition to use the law to protect plants as effectively as, for example, the RSPB has for birds.

The legal recognition of wildlife NGOs has to be distinguished from the question of exactly how they exercise any guardianship conferred upon them. When caring for children too young or mentally unable to speak for themselves, parents can call upon their own experience; they can consult any older siblings; they can seek the advice of their friends, parents and other kin; they have access to the accumulated wisdom to be found in endless autobiographies which include childhood memories. If I were given responsibility for a group of primates, I would similarly rely upon my own experience of my own needs. However, I confess that I do not know 'what it is like to be a bat'[21] and I must accept that at some point – whether with mammals, birds, amphibians or lower taxa – analogy, or induction from human experience, becomes untenable. For plant species, it seems categorically inapplicable.

Had the Sierra Club managed to persuade a fourth member of the US Supreme Court of the merits of its case, it would have been the Sierra Club, not the trees in Mineral Valley which gained standing;[22] its role as a guardian would have assumed a new significance. At some point in the future that body, or some similar wildlife organisation, will have to think very seriously about exactly what that role entails, especially in regard to determining the interests of non-human species. Twenty years' observation of the sycamore tree in my garden can be summed up very simply; if the tree can be said to have interests, these would appear to be: to become an ever bigger sycamore tree; and to have its progeny growing over an ever wider area of my garden and those of my neighbours. Forbidden by law to encroach upon the first 'ambition' by lopping any of its branches, I thwart the second by assiduously searching for and uprooting any seedlings. However, a guardian with a duty to further the tree's interests might argue, especially if an adherent to the 'selfish gene'[23] version of Darwinism, that the first interest is merely instrumental to the (superordinate) second. A deal could (and perhaps should) be struck: for each bough I remove, I agree to nurture a seedling in another area of my garden. A less jocose example might arise when the construction of a trunk road necessitates the destruction of a hazel coppice. The developers could sign an agreement to plant and maintain a greater total area of hazel coppice in other locations. Those who attribute rights to each and every tree precisely because it is an organic entity possessed of life will find the very notion of compensatory mechanisms[24] abhorrent. Whereas those who argue that it is not the individual animal or plant which bears the right to survive but the species itself (or that species' genome) may be less hostile. Regardless, the 'interest theory of rights'[25] may be capable of extension from humans to the mountain gorillas of central Africa,[26] to monk seals in the Mediterranean, to peregrine falcons in England and to any other threatened species where the need to prevent extinction is urgent. But stretching it to other more numerous animal species (especially urban pests like rats, pigeons and cockroaches) and to plants poses many problems.

Before exploring a different, more obviously ecocentric approach, it is neces-sary to consider another problem, which is closer to those encountered in earlier chapters. Let us assume that a guardian has acquired – whether by virtue of a supremely gifted intellect, imagination, empathy or insight is immaterial – a complete knowledge of the 'interests' of the species for which he is responsible. His understanding of the external factors which adversely affect those interests may nevertheless be far from total. It is no less subject to the uncertainties which beset our understanding of those threats to human health and well-being which are environmental in origin. A guardian seeking an injunction to curtail toxic emis-sions believed to have poisoned the avian inhabitants of a nature reserve still bears the onus of proof of causation. Associating the increased fragility of the eggshells of raptors with the widespread use of DDT and related insecticides was still dependent upon the advent of analytical techniques capable of measuring such low levels. The consequences of ubiquitous lead contamination are not confined to its human perpetrators. Strengthening the legal status of wildlife NGOs might lend weight to their demands for remedial action once the aetiology of harmful effects was demonstrated, but it would not remove the obligation to support their case with disinterested scientific evidence.

Botanical inhabitants of the natural world enjoy no particular immunity from environmental threats, whether anthropogenic or not. Lichen is particularly sensi-tive to sulphur dioxide; and visible damage is used as an indicator of this traditional form of air pollution. It is possible to cite other instances in which plant species are directly affected by industrial processes but where the consequences for human health and well-being may take years to be realised. Current understanding of the effects of climate change on soil biota remains speculative;[27] the consequent reduction in soil fertility could be no less than that caused directly by changes in rainfall and temperature. It is necessary to recall that it was primitive plants and bacteria which first released free oxygen into the atmosphere and thus enabled an animal kingdom to evolve. Although industrial emissions of carbon dioxide might seem numerically small when compared with the amounts continually being absorbed into marine and terrestrial 'sinks', positive feedback mechanisms in the carbon cycle raise the prospect of conditions which threaten the very survival of human (and many other vertebrate) species. Viewed in this context, the legally backed protection given to individual colonies of rare orchids smacks of senti-mental escapism. It also implies that concentrating research funds upon the lower links in the complex web of interdependency between animate and inanimate constituents of ecosystems, which modern ecology reveals to be essential to human and other forms of life, is not particularly altruistic.

Or should we simply preserve the genes?

Altruism has never been a necessary condition either for the creation of new rights or of the extension of existing ones. The abolition of slavery and the creation of a share-cropping tenancy offered landowners the prospect of an economically more efficient means of exploiting their resources. If recognition of the rights of children assists the development of cohorts of well adjusted, contented and socially respon-sible adults, this will serve to protect my pension fund and to increase my chances of securing adequate care and attention in old age or infirmity.

We may wish to protect blackbirds because we particularly enjoy their contribu-tion to the dawn chorus, and sparrows for the vicarious pleasure we derive from seeing our domestic cats pursue them. The considerable satisfaction gained from observing wild birds motivates one million people to subscribe to the RSPB. These and many others would feel a genuine sense of loss if a bird species were to become extinct or even if no longer found in the British Isles. No doubt many take the view that birds (along with other living things) have an intrinsic value which exists independently of any human need or satisfaction which they might fulfil. But the sensibilities of the individual members are unimportant compared with the organisation's achievements in advancing the cause of bird protection by sustained pressure in the political arena as well as action in the courts. Even if the Society framed its declared aims deliberately so as to avoid all trace of anthropocentrism, I do not see that this need entail a *modus operandi* markedly different from that currently employed. Its role as guardian would still be apparent, to the extent that my 'similarity' (with children) argument outlined above would still stand.

For Singer,[28] animals share with humans a capacity to experience pain which should be incorporated into any calculations of utility; for Regan,[29] it is the shared possession of consciousness which demands the inclusion of animals in the moral community. No doubt these (or other similarly non-anthropocentric) considera-tions have been influential in persuading legislatures to extend their 'humanitarian' sentiments to animals by penalising activities which, when inflicted upon humans, are deemed to be 'cruel'. Whilst there is no reason, in ethics or logic, which restricts this approach, it has a limited appeal for legislatures; effective protection is only rarely and reluctantly applied to species lacking fur or feathers.

Quite apart from a sentimental sense of loss, the extinction of any species (even insects, bacteria and slime moulds) reduces the stock of naturally evolved genes. Some may view the great diversity of living things on Earth as the clearest expres-sion of divine grace; for others it is a resource which the emergent techniques of genetic manipulation can exploit in the fight against human disease. Again, different motivations can evince equally persuasive arguments for preserving biodiversity. Some notion of the intrinsic value of higher animal taxa (such as tigers, cheetahs, humpback whales and saltwater crocodiles) may have led to their inclusion in the original Appendix I of CITES[30] prohibiting international trade in specimens (alive or dead); but subsequent versions have included a greater range of wildlife, including plants. Increasing awareness of biodiversity as a self-proliferating biological resource (offering, for example, new pharmaceutical compounds) has prompted what are utilitarian demands for a general protection, not limited to listed species, of those regions of the world (especially the tropical rain forests) with the richest and most diverse flora and fauna.

But the 'new biology' and the ever-increasing sophistication of modern techniques to produce genetically modified organisms (GMOs) have consequences which are, in two different senses, anti-environmental. First, there is the potential for unforeseen harm to ecological stability posed by alien influences (the 'rabbits in Australia' effect) which was considered at length by the Royal Commission on Environmental Pollution.[31] Another consequence has received less attention. If a species is defined by its genetic code, once we have broken that code the less important it becomes to have a reservoir of those genes existing in the natural world (whatever this phrase now means). Provided that a sufficient stock of those genes exist in laboratories (including zoological and botanical gardens) to sustain research, further erosion of natural habitats need no longer be such a cause for anthropocentric concern. Apart from the interests of tourism, why incur the considerable expense of ensuring that viable herds of black rhinoceros continue to roam the plains of Africa, when the essence of this beast – and any utility its genes may have for mankind – is preserved for posterity somewhere, be it a zoo, laboratory or database? Tropical rain forests will remain important as effective 'carbon sinks', but another argument – their indigenous genetic richness – for international cooperation to ensure their survival will become, to the layperson if not to the scientist, less cogent if and when a similar diversity can be created artificially (and far more rapidly than by natural selection).

Can the environment have rights?

Animal rights theorists have been subject to criticism from both liberal[32] and marxist[33] perspectives. Regan, and others who find intrinsic value in individual organisms, are accused of overlooking the extent to which predation (and the suffering it entails) is central to natural systems. 'Holistic' thinkers claim that an environmental ethic can and should derive from our awe of the integrity of the ecosystem and of the complex interdependencies of its animate and inanimate components.[34] Attaching no special importance to any particular species of plant or animal, or to individual members of any species, this manifestation of ecocentrism avoids the thorny question of the conflicting rights of prey and predator: a lion's right to life inescapably depends upon numerous antelopes (or other herbivores) being denied theirs.

Holistic approaches have invited the description of 'environmental fascism' when they blithely subordinate individual organisms (humans not excluded) to the aggregated utility of the entire ecosystem. If reconciling the autonomy of individuals with social order is the perennial concern of political science, then the recognition that an individual organism and the ecosystem which sustains them may both be bearers of moral consideration has become a comparably central issue in environmental philosophy.[35] This focus upon the ecosystem rather than its living constituents represents a clear shift from the narrowly anthropocentric concerns of earlier chapters. If it is meaningful to speak of the rights *of* the environment – as distinct from rights *to* an environment which can accommodate human (and other) interests – then we would have at last succeeded in our attempt to define a distinctively 'environmental right', namely one which would not fit easily into any other category.[36] But this right is one which seems destined to remain within the minds of philosophers. It is not possible to stretch further the notion of guardianship to imagine an individual or NGO with a responsibility for protecting the 'rights' of the environment as a whole.

When moral concern moves from individual species to entire ecosystems, the notion of 'stewardship' seems more appropriate than 'guardianship'. As Glacken[37] has shown, it is possible to identify a current within Judaeo-Christian thought which sees humankind's capacity to appreciate and to care for a world of infinite richness as a particular manifestation of God's grace bestowed upon the species created in his own image. White[38] was perhaps the first of a succession of writers who attribute our present plight to the other (i.e. mankind having dominion over the natural world) paradigm and to our over-eager obedience to the command: 'Be fruitful and multiply' (Genesis 9:1). More recently, other writers have used the concept of 'stewardship' as the basis for questioning the right of a landowner to waste, destroy or otherwise use 'his' land in a manner indifferent to the wider community interest.[39]

Our apprehensions of the extent of anthropogenic damage to natural systems, of the depletion of non-renewable resources, and of an increased rate of species extinction are now of such gravity that only intervention by the state seems appropriate. The more inter-related the causes and the effects, the less appropriate it is to delegate responsibility to some separate portfolio bearing the 'environmental' title and the more ecological and resource implications must permeate, if not determine, all policies. If the environment has a right, it is not to some remedy or retribution in a court, but to a permanent and prominent place in the political agenda.

The assertion that future generations have no less a claim than our own to the Earth's coal reserves, if acted upon, encourages research into alternative energy sources if not a total curb on mining. Acting *as if* the atmosphere had a right to a carbon dioxide concentration determined by non-anthropogenic processes alone might have spared us the as-yet unrealised consequences of global warming. More generally, acting as if the environment enjoys a right of non-interference, with the burden of proof of zero damage placed firmly upon the would-be polluters, is simply another expression of the precautionary principle. So seductive has the rhetoric of rights become that we are slow to recognise their metaphorical usage. Moreover, the legacy of Hohfeld is such that no duty is believed to exist independently of some correlative right. Rights have a 'dynamic'[40] quality which enables them readily to generate derivative rights and new duties appropriate to changing circumstances. But morality is not exclusively or inevitably based upon rights. Duties can exist independently of rights. It must be remembered that bear-baiting was outlawed in England many years before there was any widespread concern about the possible extinction of bears, or any other European mammalian species. Many of the statutes by which the causing of suffering to animals incurs criminal penalties predate the abolition of corporal punishment of human offenders. A belief which is still an important motivation of the animal rights movement, that other species share with humans the capacity to experience pain, was a driving force in legislation at a time when the notion of intrinsic *human* rights still provoked almost total ridicule.

But the notion of a duty not balanced by a right has too Calvinist a connotation; it does not sit easily in a culture of individual freedom and self-realisation. As an example of an intrinsic duty, Raz[41] cites the duty not to destroy a Van Gogh painting or any work of art in my possession. I am not legally disabled from burning it, and no one can be given an injunction preventing me from doing so. To deny present and future generations the pleasure of viewing the painting is a heinous act; we are all under a duty to respect the values through which others give meaning to their lives even if we do not share those values.

I suggest that there are convincing reasons for treating the environment in a similar manner. The natural beauty of animals, plants and landscapes has always been (and, one must assume, will continue to be) a primary source of aesthetic satisfaction. Even if such sentiments are not shared by all the world's inhabitants (human or otherwise), no-one can be indifferent to the environment as the sole means of satisfying our physiological needs. The graver the threats to that unique source, the more binding the duty to preserve it. To argue that it is the state which bears that duty is to make a vain attempt to deny the extent to which individual freedom in lifestyles must be circumscribed. If an environmental duty imposed upon a state can confer a right upon individuals,[42] then this right is, I suggest, no more than the privilege of reminding the state of the binding nature of the original duty. Moreover the state's duty may be one which is only fully discharged when all citizens recognise their own roles and duties (for example, to recycle paper, to minimise their reliance on fossil fuels, or to refuse to buy products made of ivory, seal skin or even tropical hardwood) in regard to environmental protection. The environment would seem to be more suited, especially given current awareness of its fragility, to a discourse of duties. If treating the environment (or any of its component parts) as if it were a rights-bearing entity assists a recognition of the corresponding duties, so be it; but it cannot make them optional or less onerous.

NOTES

CHAPTER 1

1 United Nations General Assembly, *Universal Declaration of Human Rights* (1948, New York, UNO).
2 G. Hardin, 'The Tragedy of the Commons' (1968) 162, *Science*, 1243.
3 Principle 1 of the 'Stockholm Declaration', in *The Report of the United Nations Conference on the Human Environment, 5–16 June 1972* (1973, New York) (UN doc A/Conf. 48/14/Rev.1).
4 World Commission on Environment and Development, *Our Common Future* (1987) 348.
5 'Declaration on the Human Environment', *The Report of the United Nations Conference on the Environment and Development* (1992, New York) (UN doc A/Conf. 151/26/Rev.1).
6 Joseph Raz, *The Morality of Freedom* (1986, Oxford University Press) 169.
7 See I. Brownlie, *Basic Documents on Human Rights* (3rd edn, 1992, Oxford University Press) 521.
8 For a critique of this view, see Ronald Dworkin, *Taking Rights Seriously* (1977, London, Duckworth) Chapter 4.
9 The key concept in the 'will theory' of human rights; see L. Lomasky, *Rights, Persons and the Moral Community* (1987, Oxford University Press).
10 J. Lovelock, *The Ages of Gaia: a Biography of our Living Earth* (1989, Oxford University Press).
11 W. N. Hohfeld, *Fundamental Legal Conceptions* (1919, New Haven CT, Yale University Press).
12 T. O'Riordan, *Environmentalism* (1981, Pion) 11.
13 L. Scarman, *English Law – The New Dimension* (1976, Stevens).
14 N. MacCormick, *Legal Right and Social Democracy: Essays in Legal and Political Philosophy* (1982, Oxford University Press) 143.
15 Strictly these are not 'commons' but examples of collectively owned resources where custom has restricted the right of access to those born or resident in a clearly defined area. T. O'Riordan and R. K. Turner, *An Annotated Reader in Environmental Planning and Management* (1983, Oxford, Pergamon) at 265.
16 P. R. Ehrlich and A. H. Ehrlich, *Population, Resources, Environment: Issues in Human Ecology* (1970, San Francisco CA, W. H. Freeman and Company).
17 *op. cit.* note 4, at 43.
18 s.4, Environment Act 1995.
19 Department of the Environment, *General policy and principles*, PPG1 (1992, HMS0).
20 *Great Portland Estates v. City of Westminster* [1984] 3 All ER 744.
21 J. Jowell and A. Lester, 'Beyond *Wednesbury*: substantive principles of administrative law' [1987] *Public Law*, 368–82.

22 W. Burnham, 'The constitution, capitalism and the need for rationalised regulation', in R. Goldwyn and W. Schambra (eds) *How Capitalistic is the Constitution?* (1982, Washington DC, American Enterprise Institute) quoted in P. Dunleavy and B. O'Leary, *Theories of the State* (1987, Macmillan).

23 G. L. Clark and M. J. Dear, *State Apparatus* (1984, Boston MA, Allen & Unwin).

24 Margaret Thatcher, *Keith Joseph Memorial Lecture*, edited version reproduced in the *Guardian*, 12 January 1996.

25 M. Purdue, 'Integrated pollution control in the Environmental Protection Act 1990: a coming of age of environmental law?' (1991) 54, *Modern Law Review*, 534.

26 *Davy v. Spelthorne B.C.* [1984] A.C. 262 at 276.

27 *St Helen's Smelting Co. v. Tipping* (1865) 11HL Cas 642, 11 ER 1483.

28 *Hunter and others v. Canary Wharf Ltd; Hunter and others v. London Docklands Development Corp* [1997] 2 All ER 426.

29 But, as the House was aware, Canary Wharf was sited within an 'enterprise zone' in which the normal planning procedures were suspended. Most construction went ahead with no planning consent at all; a building over 120ft in height needed the 'agreement' of the London Docklands Development Corporation.

30 *Khorasandjian v. Bush* [1993] 3 All ER 669, [1993] QB 727, [1993] 3 WLR 476, C.A.

31 *op. cit.* note 28 at 434.

32 *Cassidy v. Dunlop Rubber Co. Ltd* (1971) 11 K.I.R 311; *Wright v. Dunlop Rubber Co. Ltd* (1972) 13 K.I.R 255.

33 *Margereson and Hancock v. J. W. Roberts Ltd* [1995] 251 ENDS Report 40; J. Steele and N. Wikeley, 'Dust on the streets and liability for environmental cancers' (1997) 60, *MLR*, 265.

34 L. Arblaster, P. Hatton, E. Renvoize and M. Schweiger, *Leeds Mesothelioma Deaths 1971–87* (1990, Leeds Health Authority).

35 By 'indirect exposure' we are referring to the type of situation described in *Gunn v. Wallsend Slipway and Engineering Co. Ltd* (1989) *The Times*, 23 January, in which it was held that an employer's duty of care did not extend to the family of an employee (whose wife's fatal lung cancer was alleged to have been caused by asbestos fibres brought into the home on her husband's clothing).

36 *op. cit.* note 33.

37 The insertion of the term 'pulmonary' has been criticised as an unnecessary complication by J. Steele and N. Wikeley, *op. cit.* note 33.

38 *Budden v. BP Oil* (1980) 124 Sol Jo. 376.

39 *ibid.*

40 For a highly revisionist account, see R. M. Harrison, 'A perspective on lead pollution and health, 1972–92', *J. Roy. Soc. Health*, June 1993, 142–8.

41 Directive 78/611/EEC [1978] OJ L197/19 with its requirement that the concentration of lead in petrol should lie between 0.4 and 0.15g/litre was implemented in the UK by the Motor Fuel (Lead content of petrol) (Amendment) Regulations 1979 (SI 1979, no. 1).

42 *op. cit.* note 38.

43 J. McLoughlin and E. Bellinger, *Environmental Pollution Control* (1993, London, Graham & Trotman) 20.

44 *Cambridge Water Co. Ltd v. Eastern Counties Leather plc* [1994] 1 All ER 53.

45 'that the person who for his own purposes brings onto his land and collects and keeps there anything likely to do mischief if it escapes, must keep it in at his peril, and, if he does not do so, is *prima facie* answerable for all the damage which is the natural consequence of its escape', Blackburn, J.'s articulation of the rule in the lower court (1865) 3, *H&C*, 774.

46 *op. cit.* note 44.

47 E. Ashby and M. Anderson, *The Politics of Clean Air* (1981, Oxford University Press) Chapter 5.

48 The offence which led to the first unsuspended custodial sentence for a serious environmental crime in the UK (1992) 214, *ENDS Report*, 42.

49 *op. cit.* note 47.

50 s.80, Environmental Protection Act 1990.

51 *op. cit.* note 43 at 17.

52 'Greenpeace's expensive lessons over ICI prosecutions' (1994) 234, *ENDS Report*, 46.

53 *Burton v. Albright and Wilson* (1992) 4 LMELR 56.

54 *Manchester Corp. v. Farnworth* [1930] A.C. 171.

55 W. V. H. Rogers, *Winfield & Jolowicz on Tort* (XIIIth edition, 1989, Sweet & Maxwell) 411.

56 See *Wheeler and Another v. J. J. Saunders Ltd and Others* [1995] 2 All ER 697, in which Gibson, L. J. held: 'The Court should be slow to acquiesce in the extinction of private rights without compensation as a result of administrative decisions which could not be appealed and were difficult to challenge'.

57 s.77(2), Civil Aviation Act 1982.

58 *ibid.*

59 s.20, Land Compensation Act 1973 and the Noise Insulation Regulations 1975 (SI 1975 no. 1763).

60 s.12(1), Nuclear Installations Act 1965.

61 According to McLoughlin, this section was inserted by the Lords, after its second reading in the Commons, who accepted the representations of industry that disclosure of information on the content of discharges could be of benefit to commercial rivals, J. McLoughlin, *The Law Relating to Pollution* (1972, Manchester University Press) 35.

62 Royal Commission on Environmental Pollution, *Tenth Report: Tackling Pollution – Experience and Prospects* (Cmnd 9149, 1984, HMSO) para 2.77.

63 s. 22(5), Environmental Protection Act 1990. This appeal procedure has been invoked successfully on three occasions by major industries (National Power, Shell and ICI). The one other appeal, by PowerGen, was contested by HMIP and it involved a formal 'hearing' rather than written submissions. The recommendation of the 'person appointed' (Sir Michael Giddings, an industrialist) to allow the appeal was overturned by the Secretary of State.

64 Directive 90/313/EEC [1990] OJ L158/56.

65 Environmental Information Regulations 1992 (SI 1992, no. 324).

66 House of Lords Select Committee on the European Communities, *Freedom of Access to Information on the Environment*, 1st Report, 1996–7 Session (1996, HMSO).

67 *ibid.*, 178.

68 *Foster v. British Gas* [1990] 3 All ER 897.

69 s.31(3), Supreme Court Act 1981.

70 *R. v. Secretary of State for the Environment ex parte Rose Theatre Trust Co.* [1990] 1 QB 504.

71 O. MacIntyre and T. Mosedale, 'The rise of environmental judicial review' (1997) 6, *Environmental Policy and Practice*, 1; C. Hilson and I. Cram, 'Judicial review and environmental law – is there a coherent view of standing?' (1996) 16, *Legal Studies*, 1.

72 *R. v. HM Inspectorate of Pollution and Another, ex parte Greenpeace (no. 2)* [1994] 4 All ER 329.

73 Law Commission, *Judicial Review and Statutory Appeals* (Report no. 226, 1994, HMSO).

74 by amendment of s.31 of the Supreme Court Act 1981.

75 Lord Woolf, *Access to Justice: Final Report to the Lord Chancellor on the Civil Justice System in England and Wales* (1996, HMSO), see recommendation 240, page 321.

76 *ibid.* at 250.

77 in the chapter devoted to 'multi-party' actions, *ibid.* at 233.

78 see the article by Sir Harry (later Lord) Woolf, 'Are the judiciary environmentally myopic?' (1992) 4 *JEL* 1.
79 'Stalker' is the term applied by the British press to anyone who constantly and obsessively pursues, or attempts to communicate by letter or telephone with, another person.
80 *op. cit.* note 28 at 452, where Lord Hoffman argued that harassment was given a statutory basis by the Protection from Harassment Act 1997.
81 *Lopez Ostra v. Spain* [1995] 20 EHRR 277.
82 *Buckley v. United Kingdom* [1996] JPL 1018 at 1030.
83 *op. cit.* note 7 at 326.
84 *op. cit.* note 81 at 295.
85 *ibid.*
86 *Powell and Rayner v. United Kingdom* [1990] 12 EHRR 355.
87 *Zander v. Sweden* [1994] 18 EHRR 175.
88 *ibid.*
89 Communication no. 65/1995, *Bordes and Temeharo v. France*, UN Doc. CCPR/C/57/D/645/1995.

CHAPTER 2

1 [1977] OJ C103/1.
2 Grainne de Burca, 'The Language of Rights and European Integration' in J. Shaw and G. Moore (eds) *New Legal Dynamics of European Union* (1995, Oxford, Clarendon Press) 34.
3 L. Krämer, 'Public interest litigation in environmental matters before European courts' (1996) 8 *JEL* 1, at 13.
4 Directive 70/157/EEC [1970] OJ L42/16.
5 Directive 70/220/EEC [1970] OJ L76/1.
6 Declaration of the Council of the European Community, 22 November 1972; [1992] OJ C112/1.
7 Article 2 of the Treaty of Rome 1957 as amended by the Treaty on European Union signed at Maastricht on 7 February 1992.
8 'Sustainable development . . . which meets the needs of the present without compromising the ability of future generations to meet their own needs'. World Commission on Environment and Development, *Our Common Future* (1987) 43.
9 Rescinded by the Labour Government which took office in May 1997.
10 Article 3b, Treaty of Union.
11 Toth, A., 'The principle of subsidiarity in the Maastricht Treaty' (1992) 29, *Common Market Law Review*, 1079 at 1092.
12 Krämer reported in 1996 that the ECJ had ruled on twenty-one environmental cases since 1976; there were a further thirty cases pending, but of these twenty-two were concerned with Italian implementation of law on waste. Only one (Lappel Bank, see Chapter 8) originated in Britain, *op. cit.* note 3 at 4.
13 *R. v. Secretary of State for Transport, ex parte Factortame Ltd.* [1990] ECR I-2433, [1990] 3 CMLR 1.
14 Case C-26/62, *Van Gend en Loos v. Nederlandse Administratie der Belastingen* [1963] ECR 1.
15 Case C-41/74, *Van Duyn v. Home Office* [1974] ECR 1337, [1975] I CMLR 1.
16 Case C-148/78, *Pubblico Ministero v. Ratti*, [1979] ECR 1629.
17 Case C-8/81, *Becker v. Finanzamt Munster-Innenstadt* [1982] ECR 53, [1982] I CMLR 499.
18 *ibid.* at 71.
19 L. Krämer, 'The implementation of Community Environmental Directives within member states: some implications of the direct effect doctrine' (1991) 3 *JEL* 39.

20 Case C-361/88, *Commission of the European Community v. Germany* [1991] ECR I-2567 at 2601, quoted in J. Jans 'Legal Protection in European Environmental Law' (1993) 2, *European Environmental Law Review*, 151.

21 Directive 80/779/EEC [1980] OJ L229/30. The practical implications of the 'direct effect' of this Directive are discussed at length in Chapter 4.

22 L. Krämer, 'Direct effect of EC environmental law', in H. Somsen, *Protecting the European Environment: Enforcing EC Environmental Law* (1996, Blackstone).

23 *ibid.* at 101.

24 *ibid.* at 108.

25 *ibid.* at 110.

26 A. Geddes, *Protection of Individual Rights under EC Law* (1995, Butterworth) 78.

27 Directive 90/313/EEC [1990] OJ L158/56.

28 Directive 85/337/EEC [1985] OJ L175/40.

29 *R. v. Legal Aid Area no. 8 (Northern) ex parte Florence Emily Sendall* [1993] Env LR 167.

30 A. Geddes, '*Locus Standi* and EEC environmental measures' (1992) 4 *JEL* 29 at 38.

31 Directive 76/464/EEC [1976] OJ L129/33.

32 *op. cit.* note 30 at 38.

33 *op. cit.* note 19 at 42. The ECJ ruling, which contrary to both Geddes and Krämer, held a daughter of this Directive to fail the test of direct effect is discussed in Chapter 5.

34 *op. cit.* note 19 at 47.

35 Directive 82/501/EEC [1982] OJ L230/1.

36 *op. cit.* note 19 at 48.

37 Case C-33/76, *Rewe-Zentralfinanz v. Landwirtschaftskammer* [1976] ECR 1989. The principle of state liability for breaches of Community law was articulated in an even earlier case (Case C-39/72, *Commission v. Italy* [1973] ECR 101).

38 *Twyford Parish Council and others v. Secretary of State for Transport* (1992) 4 *JEL* 273.

39 See J. Alder, 'Environmental impact assessment – the inadequacies of English law' (1993) 5 *JEL* 203.

40 See Eckard Rehbinder, '*Locus standi*, Community law and the case for harmonization' in H. Somsen, *op. cit.* note 22.

41 *op. cit.* note 26 at 80.

42 Directive 91/492/EEC [1991] OJ L268/1; Directive 91/493/EEC [1991] OJ L268/15.

43 Directive 83/129/EEC [1983] OJ L91/30.

44 Directive 79/409/EEC [1979] OJ L103/1.

45 *op. cit.* note 19 at 45–6.

46 *op. cit.* note 3 at 15.

47 There is perhaps a more convincing argument against the direct effect of the Birds Directive. The interests of persons most affected by 79/409/EEC – farmers, hunters, collectors and dealers in birds and their eggs – are not improved by its implementation; the privileges (for example, to take eggs) which they would otherwise enjoy are removed and replaced by obligations *not* to kill birds, destroy nests, collect or trade in eggs, etc. Thus it might be argued that the relevant articles of this Directive reduced the liberties of the individual and therefore, on the principle articulated in *Kolpinghuis* (Case C-80/86, *Kolpinghuis Nijmegen* [1987] ECR 3969 at 3982), cannot have direct effect.

48 House of Lords Select Committee on the European Communities, *Implementation and Enforcement of Environmental Legislation*, 9th Report, 1991–2 session).

49 Law Commission, *Administrative Law: Judicial Review and Statutory Appeals* (Law Commission Report no. 226, 1994) para. 12.18.

50 Case T-219/95R, *Danielsson and others v. Commission of the European Communities*, Court of [1995] ECR II-305.

51 Royal Commission on Environmental Pollution, *Ninth Report: Lead in the Environment* (1983, HMSO, Cmnd 8852) para. 7.115.

52 Case C-6/90, *Francovich v. Italian Republic* [1992] ECR I-5357.

53 *op. cit.* note 37.

54 *op. cit.* note 52, paras 39–40.

55 J. Steiner, 'From direct effects to *Francovich*: shifting means of enforcement of Community law' (1993) 18, *European Law Review*, 3; M. Ross, 'Beyond *Francovich*' (1993) 56, *Modern Law Review*, 55.

56 Lord Slynn, 'The European Community and the environment' (1993) 5 *JEL* 261.

57 J. Lefevere, 'State Liability for Breaches of Community Law' [1996] *European Environmental Law Review*, 237.

58 Joined Cases C-46/93 and C-48/93, *Brasserie du Pêcheur SA v. Germany* and *R. v. Secretary of State for Transport ex parte Factortame Ltd and Others*, [1996] ECR I-1209, [1996] 1 CHLR 889.

59 J. Coppel, *Individual Enforcement of Community Law: The Future of the Francovich Remedy*, EUI Working Paper LAW no. 93/6 (1993, Florence, European University Institute) quoted by H. Somsen, '*Francovich* and its application to EC environmental law', *op. cit.* note 22 at 146.

60 F. Schockweiler, 'Die Haftung der EG-Mitgliedstaaten gegenüber dem einzelnen bei Verletzung des Gemeinschaftsrechts' (1993) 23, *Europarecht*, 356, translated and quoted by H. Somsen, '*Francovich* and its application to EC environmental law', *op. cit.* note 22 at 145.

61 J. Holder, 'A dead end for direct effect?' (1996) 8 *JEL* 322 at 330.

62 Case C-106/89, *Marleasing SA v. La Comercial Internacional de Alimentacion SA* [1990] ECR I-4135, [1992] I CMLR 305.

63 *ibid.*

64 P. Pescatore, 'The doctrine of "direct effect": an infant disease of Community law' (1983) 8, *European Environmental Law Review*, 155.

65 Greenpeace did successfully argue the indirect effect of a EURATOM directive in their judicial review of the THORP authorisations (a case better known for the breakthrough on standing, see Chapter 6 below).

66 *R. v. Secretary of State for Trade and Industry, ex parte Duddridge* (1995) 7 *JEL* 224.

67 Case C-168/95, *Criminal Proceedings against Luciano Arcaro* [1996] ECR I-4705.

68 *Wychavon District Council v. Secretary of State for the Environment and Velcourt Limited* (1994) 6 *JEL* 351.

69 *ibid.* at 358.

70 *R. v. North Yorkshire County Council ex parte Brown* (1997) 265, *ENDS Report*, 42.

71 Case C-72/95, *Aanemersberdijf PK Kraajeveld BV and others v. Gedeputeerde Staten van Zuid Holland* (1997) 264, *ENDS Report*, 44.

72 *op. cit.* note 66.

73 T. O'Riordan and J. Cameron (eds) *Interpreting the Precautionary Principle* (1994, London, Earthscan) 17.

74 *op. cit.* note 66 at 229.

75 *op. cit.* note 73 at 237.

76 Compare this attitude with that of Geddes and Krämer who attach a special importance to those environmental directives which have, as a subordinate aim, the protection of human health.

77 The principle that environmental damage should be addressed at source Art. 130r(2) was part of the justification of the ECJ in upholding a Belgian prohibition of the import of waste from other member states Case C-2/90, *Commission v. Belgium* [1993] 1 CMLR 365.

78 Case C-152/84, *Marshall v. Southampton Area Health Authority* [1986] ECR 723, [1986] 1 CMLR 688.
79 Directive 76/207/EEC [1976] OJ L39/40.
80 Council on Tribunals, *Annual Report of the Council on Tribunals for 1995/96* (1996, HMSO) 104.

CHAPTER 3

1 J. B. Cullingworth, *Town and Country Planning in Britain*, 10th edn (1988, Unwin Hyman) 16.
2 Minister without Portfolio, *Lifting the Burden* (1985, HMSO, Cmnd 9571).
3 Town and Country Planning (Assessment of Environmental Effects) Order 1988 (SI 1988 no. 1199).
4 Department of the Environment, *Sustainable Development: the UK Strategy* (Cmnd 2426, HMSO, 1994).
5 Ministry of Housing and Local Government, *People and Planning* ('the Skeffington Report', 1969, HMSO).
6 P. McAuslan, *The Ideologies of Planning Law* (1980, Pergamon) 271.
7 *ibid.*
8 Town and Country Planning (Amendment) Act 1972, s.3(1).
9 Town and Country Planning Act 1990, s.20(6).
10 *ibid.*, s.35(6). The position with regard to local plans is less straightforward. If the Secretary of State calls in a local plan for his approval, he *must* consider any objections made in the prescribed manner, unless the planning authority has already done so prior to 'call in', s.45(3). The decision to 'call in' is at his discretion.
11 The Planning Inspectorate Executive Agency, *Annual Report and Accounts for the Year Ended 31 March 1996* (1996, HMSO) para 3.7.
12 *Turner v. Secretary of State for the Environment* (1973) 28 P&CR 123.
13 It is now generally accepted that the legality of decisions taken by local planning authorities (but not the Secretary of State) may be challenged by judicial review in the High Court, *R. v. Hillingdon LBC ex parte Royco Homes Ltd* [1974] QB 720, [1974] 2 All ER 643.
14 *Ryeford Homes Ltd v. Sevenoaks District Council* [1990] JPL 36.
15 *Wheeler and Another v. J. J. Saunders Ltd and Others* [1995] 2 All ER 697, which effectively reversed the ruling in *Gillingham BC v. Medway (Chatham) Dock Company* [1992] Env LR 98 for a discussion of the latter, see C. Crawford (1992) 4 *JEL* 262.
16 s.55(1), Town and Country Planning Act 1990.
17 Town and Country Planning (General Permitted Development) Order 1995 (SI 1995 no. 418).
18 Town and Country Planning (Use Classes) Order 1987 (SI 1987 no. 764).
19 *Westminster City Council v. Great Portland Estates plc* [1985] 1 AC 661 at 670.
20 *J. A. Pye (Oxford) Estates Ltd v. West Oxfordshire District Council* [1982] JPL 577.
21 Department of the Environment, *Development Control – Policy and Practice* (Circular 22/80, HMSO, 1980).
22 Department of the Environment, *Development and Employment* (Circular 14/85, HMSO, 1985).
23 *Cranford Hall Parking Ltd v. Secretary of State for the Environment* [1989] JPL 169.
24 Highways Procedures Rules 1976 (SI 1976, no. 721), rules 6–7, replaced by the Highways (Inquiries Procedure) Rules 1994 (SI 1994, no. 3263).
25 *Bushell v. Secretary of State for the Environment* [1981] AC 75, [1980] 2 All ER 608, [1980] JPL 458.

26 Barbara Bryant, *Twyford Down: Roads, Campaigning and Environmental Law* (1996, E & F Spon) 38.

27 A Roman road, and the two medieval tracks known locally as the 'Dongas' – a name subsequently adopted and made famous by the extra-legal protesters.

28 Ancient Monuments and Archaeological Areas Act 1979.

29 *op. cit.* note 26 at 103.

30 s.85, National Parks and Access to the Countryside Act 1949.

31 Directive 85/337/EEC [1985] OJ L175/40.

32 Highways (Assessment of Environmental Effects) Regulations 1988 (SI 1988, no. 1241).

33 A statutory review under Schedule 2 of the Highways Act 1980.

34 *Twyford Parish Council and Others v. Secretary of State for Transport* (1992) 4 *JEL* 273.

35 Case 8/81, *Becker v. Finanzamt Munster-Innenstadt* [1982] ECR 53, [1982] 1 CMLR 499.

36 *op. cit.* note 34 at 279.

37 *ibid.*

38 *op. cit.* note 33, para. 3(b).

39 J. Alder, 'Environmental impact assessment – the inadequacies of English law' (1993) 5 *JEL* 203.

40 *op. cit.* note 26.

41 *op. cit.* note 26 at 152.

42 See Peter Kunzlik, 'The lawyer's assessment', in Bryant, *op. cit.* note 26.

43 Department of the Environment, *Report into the Inquiry of two appeals by National Smokeless Fuels Ltd* (File nos: APP/A4520/A/87/068722 & 075692) (1988, DOE).

44 Created in 1863, HM Alkali Inspectorate was the specialist agency of central government which enforced the statutory obligation to use 'best practicable means' to limit discharges from major sources of industrial air pollution. In 1987, its successor (which was then part of the Health and Safety Executive) was absorbed into HM Inspectorate of Pollution, and this body in turn became one of the principal constituents of the Environment Agency in 1996.

45 Health and Safety Executive, *Industrial Air Pollution 1976* (1978, London).

46 The 'Convention on Long-range Transboundary Air Pollution' was drawn up under the auspices of the United Nations' Economic Commission for Europe (UNECE). The 'Geneva Convention' was signed by the European Community in November 1979; it was ratified in 1981 by Council Decision 81/462/EEC [1981] OJ L171/11 and came into force in 1983. A subsequent protocol (Helsinki) of the Convention required signatories to reduce their SO_2 emissions by 30 per cent by 1993, taking 1980 as the baseline. The European Commission represented the Community and in 1985 signed the protocol along with most member states (but not the UK or Eire). Subsequent UK policy, especially in regard to the fitting of flue gas desulphurisation to coal-fired power plants, meant that the UK became a *de facto* member of the '30 per cent Club'.

47 *op. cit.* note 43.

48 Department of the Environment, *Decision Letter: Appeals by National Smokeless Fuels Ltd* (ref. nos: APP/A4520/A/87/068722 & 075692) (1988, Northern Regional Office).

49 Under powers now contained in s.288 of the Town and Country Planning Act 1990, see Table 3.1.

50 *National Smokeless Fuels Ltd and Another v. the Secretary of State for the Environment*, unreported, 1989.

51 s.37 of the 1971 Act; s.78 of the 1990 Act.

52 Department of the Environment, *Report of the Re-opened Inquiry into an Appeal by National Smokeless Fuels and Coal Products Ltd* (file no.: APP/A4520/A/87/075692) (1990, DOE Northern Regional Office).

53 Townsend, P., Phillimore, P. and Beattie, A., *Health and Deprivation: Inequalities and the North* (1988, Croom Helm).

54 World Health Organisation, *Air Quality Criteria and Guides for Urban Air Pollutants: Technical Report Series no. 506* (1972, Genevea, WHO).
55 Directive 80/779/EEC [1980] OJ L229/30.
56 Directive 84/360/EEC [1984] OJ L188/20.
57 Directive 88/609/EEC [1988] OJ L336/1.
58 *op. cit.* note 56.
59 Department of the Environment, *Decision Letter: Appeal by National Smokeless Fuels Ltd and Coal Products Ltd* (ref. no.: APP/A4520/A/87/075692) (1991, DOE Northern Regional Office).
60 *ibid.*
61 Environmental Protection Act 1990 (Commencement no. 4) Order 1990 (SI 1990, no. 2635).
62 *op. cit.* note 56.
63 Case C-14/83, *Von Colson and Kamann v. Land Nordrhein-Westfalen* [1984] ECR 1891.
64 Case C-106/89, *Marleasing v. La Comercial Internacional de Alimentacion* [1990] ECR I-4135.
65 Department of the Environment, 'Planning and clean air' (1972, draft circular).
66 See note 44 above.
67 Royal Commission on Environmental Pollution, *Air Pollution Control: An Integrated Approach. Fifth Report* (Cmnd 6371, 1976, HMSO).
68 Department of the Environment, *Planning Policy Guidance: Planning and Pollution Control* (PPG23, 1994, HMSO) para. 3.23.
69 *Gateshead MBC v. Secretary of State for the Environment* [1995] JPL 432. Reference has already been made to this particular inquiry (Chapter 2, Box 2.3) in regard to a third-party objector's attempt to invoke 'direct effect' to secure legal aid to assist her participation.
70 *ibid.* at 439.

CHAPTER 4

1 The Labour Party, *In trust for tomorrow: Report of the Labour Party Policy Commission on the Environment* (1994).
2 *St Helen's Smelting v. Tipping* (1865) 11 H.L. Cas. 642.
3 *Sturges v. Bridgman* (1879) 11 ChD 856.
4 *ibid.*
5 *Rylands v. Fletcher* (1866) L.R. 1 Ex. 265, affirmed (1868) L.R. 3 H.L. 330.
6 s.82, Environmental Protection Act 1990.
7 s.7 of the Alkali Act 1906 until superseded by s.5 of the Health and Safety at Work Act 1974.
8 E. Ashby and M. Anderson, *The Politics of Clean Air* (1981, Oxford University Press).
9 Department of the Environment, *108th Annual Report on Alkali, etc. Works 1971* (1972, HMSO).
10 *ibid.*
11 The need to apply the 'best practicable environmental option' for processes polluting more than one environmental medium does not apply to local authority controlled processes (s. 7(7) of the 1990 Act).
12 See especially Framework Directive 84/360/EEC [1984] OJ L188/20 and its daughter directives.
13 s.3(5); for example, national targets for sulphur emissions under the 'Large Combustion Plant Directive' (88/609/EEC [1988] OJ L336/1) could entail conditions upon the authorisation of each fossil-fuelled power plant.

14 Royal Commission on Environmental Pollution, *Tenth Report: Tackling Pollution – Experience and Prospects* (Cmnd 9149, 1984, HMSO).

15 *ibid.*, para. 2.78.

16 s.28(7), Health and Safety at Work Act 1974.

17 Directive 90/313/EEC [1990] OJ L158/56.

18 1990 Act, s.6(6).

19 Environmental Protection (Applications, Appeals and Registers) (Amendment) Regulations 1996 (SI 1996, no. 667).

20 Her Majesty's Inspectorate of Pollution *1995–6 Annual Report* (1996, HMSO) 53.

21 A relatively cheap fuel consisting of bituminous material imported from Venezuela, [1995] 240, *ENDS Report*, 8.

22 'Renderer loses High Court case but still thwarts air pollution rules' [1996] 254, *ENDS Report*, 7.

23 'HMIP blocks major gas project in BATNEEC test case' [1995] 245, *ENDS Report*, 13.

24 Especially the suspected carcinogens – dioxins, furans and polychlorinated bi-phenyls (PCB).

25 House of Commons Environment Committee, *The Burning of Secondary Liquid Fuel in Cement Kilns*, Second Report, session 1994–5 (1995, HMSO).

26 Redland Aggregates Ltd, *Assessment of the Use of Solvent Derived Fuels (SDF): Whitwell Works* (1995) 4.

27 Figures quoted by a witness, 7 June 1995, *op. cit.* note 25.

28 s.34 of the 1990 Act, see Chapter 7 below.

29 Directive 94/67/EEC [1994] OJ L365/34.

30 *op. cit.* note 21.

31 'Cement kilns and hazardous waste – the debate heats up' [1994] 233, *ENDS Report*, 14.

32 Ministry of Housing and Local Government, *103rd Annual Report on Alkali, etc. Works 1966* (1967, London, HMSO) 9.

33 Directive 80/779/EEC [1980] OJ L229/30.

34 Air Quality Standards Regulations 1989 (SI 1989, no. 317).

35 Department of the Environment, *Digest of Environmental Protection and Water Statistics no. 14 1991* (1992, London, HMSO) 6.

36 Now under s.18, Clean Air Act 1993 (which consolidates and replaces the Acts of 1956 and 1968).

37 *op. cit.* note 8.

38 R. Bhopal, S. Moffat, P. Phillimore and C. Foy, *The Monkton Coking Works Study* (1992, University of Newcastle upon Tyne).

39 D. W. Dockery, C. A. Pope III, X. Xu, J. D. Spengler, J. H. Ware, M. E. Fay, B. G. Ferris Jnr and F. E. Speizer, 'An association between air pollution and mortality in six US cities', 329 (24) (1993) *New England Journal of Medicine*, 1754.

40 One micron represents one millionth of a metre.

41 Case C6/90, *Francovich v. Italian Republic* [1992] ECR I-5357.

42 s.19, Clean Air Act 1993.

43 Department of the Environment, *Digest of Environmental Statistics no. 18 1996* (1996, London, HMSO) 25.

44 *ibid.*, 26.

45 *R. v. London Borough of Greenwich ex parte Jack Williams (an infant) by Kathleen Williams (his mother and next friend) and Others* [1997] JPL 62.

46 Directive 92/72/EEC [1992] OJ L297/1.

47 Most exceedences (namely thirty-nine occasions during eight days) of the information threshold in 1994 were recorded at the rural (Sibton in East Anglia) site. The warning threshold was exceeded at none of the twenty-nine sites where ozone monitoring was effective in 1994; *op. cit.* note 43, at 39.

48 Department of the Environment, *Air Quality: Meeting the Challenge. The Government's Strategic Policies for Air Quality Management* (1995, London).

49 Department of the Environment, *Improving Air Quality: A Discussion Paper on Air Quality Standards and Management* (1994, London) para. 3.6.

50 s.7(2) of the 1990 Act.

51 Department of the Environment, *Local Authority Circular on Air Quality and Traffic Management: Consultation Paper* (1995, DOE).

52 s.39 of the Environment Act 1995.

53 Clean Air Act 1970, s.109(b)(1–2) 42 U.S.C. 7409(b) (1–2).

54 M. B. Mackay, 'Environmental rights and the US system of protection: why the US Environmental Protection Agency is not a rights-based administrative agency' (1994) 26, *Environment and Planning A*, 1761.

55 R. B. Stewart, 'Economics, environment and the limits of legal control' (1985) 9, *Harvard Environmental Law Review*, 1.

56 'Clean power company sells right to pollute', *Guardian*, 13 May 1992.

57 Secretary of State for the Environment *et al.*, *This Common Inheritance: Britain's Environmental Strategy*, (Cmnd 1200 1990, HMSO).

58 [1995] 244, *ENDS Report*, 39.

59 Directive 88/609/EEC [1988] OJ L336/1.

60 [1994] 230, *ENDS Report*, 36.

61 [1995] 248, *ENDS Report*, 11.

62 A plant's allocated emission quota would be taken into account in the setting of conditions of the authorisation. See the 'Large Combustion Plant Directive' *op. cit.* note 59.

63 See Critical Loads Advisory Group, *Critical Loads of Acidity in the United Kingdom* (1994, Institute of Terrestrial Ecology).

64 *op. cit.* note 21.

65 'Acid deposition study last straw for SO_2 trading' [1996] 253, *ENDS Report*, 12.

66 [1994] 232, *ENDS Report*, 6.

67 According to Scherer, the fund established by the Dutch Air Pollution Act of 1970 compensates for damage arising from 'sudden' air pollution rather than the 'chronic pollution' with which our discussion is most concerned; see J. Scherer, 'Restoration and prevention of environmental damage through joint compensation systems' [1994] *European Environmental Law Review*, 161.

68 Pollution-related Health Damage Compensation Law, Act No 111 of 5 October 1973, see Scherer, *ibid.*

69 Minimata is a fishing community in Japan. In the 1960s fishermen and their families began to suffer severe neurotoxic effects. These were later attributed to the mercury poisoning acquired by eating fish caught in waters contaminated by methyl mercury discharged from a factory producing vinyl chloride. Some 2,000 people were affected, including 43 fatalities and 700 permanently disabled.

70 R. Dworkin, *Taking Rights Seriously* (1977, London, Duckworth) 90.

CHAPTER 5

1 The classic economic analysis of such resources took as its subject 'the fishery'. H. S. Gordon, 'The economic theory of a common property resource' (1954) 62, *Journal of Political Economy*, 124.

2 In two 'cod wars', Iceland's small but determined navy successfully defended its claims to sole access to the fisheries within a 200-mile zone. The prospect of the loss of exclusive access to this rich resource (the basis of the Icelandic economy) has proved a strong disincentive to Iceland's applying for membership of the European Union.

The 200 nautical mile 'exclusive economic zone' established by United Nations Convention on the Law of the Sea, or any other extension of the concept of territorial waters, is antithetical to the notion of a 'commons'.

3 *Young & Co. v. Bankier Distillery Co.* [1893] AC 691.

4 R. H. Coase, 'The problem of social cost' (1960) 3, *Journal of Law and Economics*, 1.

5 G. Calabresi and A. Melamed, 'Property rules, liability rules and inalienability: one view of the cathedral' (1972) 85, *Harvard Law Review*, 1089 at 1097.

6 *Hughes v. Dwr Cymru Cyf* (Llangefni County Court) (1995) 247, *ENDS Report* 42 implies the defence of statutory authority applies once consent to discharge is complied with. This ruling, if not overturned in the Court of Appeal, is contrary to what has been seen hitherto as a definitive ruling in *Pride of Derby etc. Association Ltd v. British Celanese* [1953] 1, Ch. 149.

7 See W. Howarth, '"Poisonous, noxious or polluting": contrasting approaches to environmental regulation' (1993) 56 *Modern Law Review*, 171 at 180.

8 *National Rivers Authority v. Shell UK* [1990] Water Law 40.

9 Directive 80/68/EEC [1980] OJ L20/43.

10 Case C-131/88, *Commission v. Federal Republic of Germany* [1991] ECR I-825, para. 7 of the judgment.

11 *ibid.* at 850.

12 L. Krämer, 'The implementation of Community environmental directives within member states: some implications of the direct effect doctrine' (1991) 3 *JEL* 39 at 45.

13 *op. cit.* note 9.

14 (1866) L.R. 1 Ex. 265, affirmed (1868) L.R. 3 H.L. 330. at 280.

15 D. Wilkinson, '*Cambridge Water Company v. Eastern Counties Leather plc*: diluting liability for continuing escapes' (1994) 57 (5) *The Modern Law Review*, 799.

16 Directive 76/464/EEC [1976] OJ L129/33.

17 *op. cit.* note 9.

18 Directive 83/513/EEC [1983] OJ L291/1.

19 C-168/95, *Criminal Proceedings against Luciano Arcaro* [1996] 1 ECR 4705.

20 *ibid.* at 4713.

21 Directive 76/160/EEC [1990] OJ L30/1.

22 Nigel Haigh, *EEC Environmental Policy and Britain*, 2nd edn (1987, Longman) 65.

23 'Windsurfer sues over virus in sea', *Guardian*, 17 September 1993. I am unaware of this case or any other concluding with a plaintiff successful in securing damages in respect of health detriment attributed to swimming in contaminated waters.

24 'Surfers face viral infection risks from sewage pollution' (1993) 235, *ENDS Report*, 9.

25 D. Kay, J. M. Fleischer, R. L. Salmon, F. Jones, M. D. Dwyer, A. F. Godfree, Z. Zelanauch-Jacquotte and R. Shore, 'Predicting likelihood of gastroenteritis from sea bathing: results from randomised exposure' (1994) 344, *The Lancet*, 905.

26 Case C-56/90, *Commission of the European Communities v. United Kingdom of Great Britain and Northern Ireland* (1994) 6 *JEL* 125.

27 *ibid.* at 134.

28 *R. v. National Rivers Authority, ex parte Morton* (1994–5) 1 EJRB 56.

29 Bathing Waters (Classification) Regulations 1991 (SI 1991, no. 1597).

30 *op. cit.* note 28 at 57.

31 Directive 91/271/EEC [1991] OJ L135/40.

32 Urban Waste Water Treatment (England and Wales) Regulations 1994 (SI 1994, no. 2841).

33 *R. v. Secretary of State for the Environment, ex parte Kingston upon Hull City Council and R. v. Secretary of State for the Environment, ex parte Bristol City Council and Woodspring District Council* (1996) 8 *JEL* 336.

34 Case C-44/95, *R. v. Secretary of State for the Environment, ex parte RSPB* (1997) 9 *JEL* 139.

35 *R. v. Carrick District Council ex parte Shelley* (1996) 255, *ENDS Report*, 48.

36 s.80, Environmental Protection Act 1990.

37 This defence, s.31(5) of the Control of Pollution Act 1974, was rarely cited in court according to W. Haworth, 'Water pollution: improving the legal controls' (1989) 1 *JEL* 25.

38 The Nitrate Sensitive Areas (Designation) Order 1990 (SI 1990, no. 1013).

39 Case C-337/89, *Commission of the European Communities v. United Kingdom of Great Britain and Northern Ireland* [1992] 1 ECR 6103.

40 Directive 80/778/EEC [1980] OJ L229/11, implemented in the UK by the Water Supply (Water Quality) Regulations 1989 (SI 1989, no. 1147).

41 *R. v. Secretary of State for the Environment, ex parte Friends of the Earth Limited* (1995) 7 *JEL* 80.
      It should be noted that the standing of FOE to bring this judicial review, six months after Greenpeace's first challenge of THORP's discharges, was not challenged by the Secretary of State.

42 s.19(1)b, Water Industry Act 1991.

43 *R. v. Secretary of State for the Environment, ex parte Friends of the Earth and Another* (Court of Appeal, 25 May 1995; subsequent references are to the pages of the transcript of the judgement).

44 s.59(3)(b), Town and Country Planning Act 1990. See Chapter 3 for a discussion of some of the consultation procedures which apply to development other than those approved by a 'special development order'.

45 *op. cit.* note 43 at 25.

46 *op. cit.* note 9.

47 *op. cit.* note 9, para. 6.

48 *op. cit.* note 9, para. 8.

49 *op. cit.* note 43 at 47.

50 *ibid.*

51 *op. cit.* note 43 at 25.

52 Joined Cases C-46/93 and C-48/93, *Brasserie du Pêcheur SA v. Germany and R. v. Secretary of State for Transport ex parte Factortame Ltd and Others*, [1996] ECR I-1029, [1996] 1 CMLR 889.

53 'Friends of the Earth loses drinking water case' (1995) 245, *ENDS Report*, 44.

54 *op. cit.* note 9.

55 N. F. Gray, *Drinking Water Quality: Problems and Solutions* (1994, Chichester, Wiley)

56 *op. cit.* note 26.

57 World Health Organisation, *Revision of the WHO Guidelines for Drinking Water Quality* (1993, Geneva, WHO)

58 R. Macrory, *Water Law: Principles and Practice* (1985, London, Longman) 105.

59 Royal Commission on Environmental Pollution, *Ninth Report: Lead in the Environment* (Cmnd 8852 1983, HMSO) para. 8.23.

60 *Budden v. BP Oil* (1980) 124 Sol Jo. 376.

61 *ibid.*

62 *Read v. Croydon* [1938] 4 All ER 631.

63 *Cryptosporidium* has been responsible for a number of recent outbreaks of gastrointestinal disorders. No reference is made to this protozoon in the Drinking Water Directive, *op. cit.* note 40.

64 M. Purdue, 'The possible will take a long while – enforcing compliance with the Drinking Water Directive' (1995) 7 *JEL* 80 at 97.

65 Dame Barbara Clayton (Chairperson) *Water Pollution at Lowermoor, North Cornwall: Second Report of the Lowermoor Incident Health Advisory Group* (1991, Department of Health).

66 Charles Pugh and Martyn Day, *Toxic Torts* (1992, London, Cameron May) 130.

67 Which implements a directive on product liability, 85/374/EEC [1985] OJ L210/29.

68 *Donoghue (or M'Alister) v. Stevenson* [1932] AC 562.

69 Ingestion of industrial rapeseed oil, denatured with aniline and further treated with other oils of plant and animal origin, led to fever, respiratory distress, nausea, vomiting and myalgia. B. Hobbs and D. Roberts, *Food Poisoning and Food Hygiene*, 6th edn (1993, London, Arnold).

70 Severn Trent subsequently failed to secure a conviction against the solvent recycling business for an illegal entry to the sewers (1995) 249, *ENDS Report*, 41.

71 s.70, Water Industry Act 1991; see 'Severn Trent convicted for Worcester drinking water incident' (1995) 243, *ENDS Report*, 45.

72 s.45, Water Industry Act 1991.

73 *ibid.*, s.106.

74 A. Herbert and E. Kempson, *Water Debt and Disconnection* (1995, Policy Studies Institute).

75 Water Industry Act 1991, s.2.

76 *op. cit.* note 15.

CHAPTER 6

1 International Commission on Radiological Protection, *Recommendations of the International Commission on Radiological Protection* (1977, Pergamon).

2 *Hope v. British Nuclear Fuels Ltd; Reay v. BNFL, The Guardian Law Report*, 15 October 1993.

3 *R. v. HM Inspectorate of Pollution and Another, ex parte Greenpeace (no. 2)* [1994] 4 All ER 329.

4 s.16(1) of the Nuclear Installation Act 1965, as amended by s.27 of the Energy Act 1983.

5 *Re Friends of the Earth* [1988] JPL 93.

6 *ibid.* at 98.

7 See C. E. Miller, 'Economics v. pragmatics: the control of radioactive wastes' (1990) 2 *JEL* 65.

8 *Merlin v. British Nuclear Fuels plc* (1991) 3 *JEL* 122.

9 *Convention on Civil Liability for Nuclear Damage* (Cmnd 2333) signed at Vienna on 21 May 1963. The need to implement this convention in the UK was one of the reasons for replacing the Nuclear Installations (Licensing and Insurance) Act 1959 with the 1965 Act.

10 *ibid.* at Article I(k)(ii).

11 *op. cit.* note 5.

12 *op. cit.* note 8 at 130.

13 *Blue Circle Industries plc v. Ministry of Defence, The Times*, 11 December 1996.

14 s.12(1).

15 *op. cit.* note 2.

16 *op. cit.* note 8.

17 D. R. Miller, 'Courtroom science and standards of proof', *The Lancet*, 28 November 1987, 1283–4.

18 A. S. Levin, 'Science in court', *The Lancet*, 26 December 1987, 1529.

19 A. B. Hill, 'The environment and disease: association or causation' (1965) 58, *Proceedings of the Royal Society of Medicine*, 1217–19.

20 M. J. Gardner, M. P. Snee, A. J. Hall, C. A. Powell, S. Downes and J. D. Terrell, 'Results of a case-control study of leukaemia and lymphoma among young people near Sellafield nuclear plant in West Cumbria' (1990) 300, *British Medical Journal*, 423–9.

21 Independent Advisory Group, *Investigation of the Possible Increased Incidence of Cancer in West Cumbria* (1984, HMSO).

22 Committee on Medical Aspects of Radiation in the Environment, *Second Report. Investigation of the Possible Increased Incidence of Leukaemia in Young People near the Dounreay Nuclear Establishment, Caithness, Scotland* (1988, HMSO).

23 Committee on Medical Aspects of Radiation in the Environment, *Third Report. Report on the Incidence of Childhood Cancer in the West Berkshire and North Hampshire Area, in which are Situated the Atomic Weapons Research Establishment, Aldermaston and the Royal Ordnance Factory, Burghfield* (1989, HMSO).

24 L. J. Kinlen, K. Clarke and C. Hudson, 'Evidence from population mixing in British New Towns 1946–85 of an infective basis for childhood leukaemia' (1990) 336, *The Lancet*, 577–82.

25 L. J. Kinlen, F. O'Brien, K. Clarke, A. Balkwill and F. Matthews, 'Rural population mixing and childhood leukaemia: effects of the North Sea oil industry in Scotland, including the area near Dounreay nuclear site' (1993) 306, *British Medical Journal*, 743–8.

26 L. J. Kinlen, K. Clarke, A. Balkwill and F. Matthews, 'Paternal preconceptional radiation exposure in the nuclear industry and leukaemia and non-Hodgkin's lymphoma in young people in Scotland' (1993) 306, *British Medical Journal*, 1153–8.

27 L. J. Kinlen, 'Can paternal preconceptional radiation account for the increase in leukaemia and non-Hodgkin's lymphoma in Seascale?' (1993) 306, *British Medical Journal*, 1718–21.

28 *op. cit.* note 21.

29 L. Parker, A. W. Craft, J. Smith, H. Dickinson, R. Wakeford and K. Binks, 'Geographical distribution of preconceptional radiation doses to fathers employed at Sellafield nuclear installation, West Cumbria' (1993) 307, *British Medical Journal*, 966–71.

30 J. R. McLaughlin, W. D. King, T. W. Anderson, A. E. Clarke and J. P. Ashmore, 'Paternal radiation exposure and leukaemia in offspring: the Ontario case-control study' (1993) 307, *British Medical Journal*, 959–66.

31 *op. cit.* note 26.

32 Y. Yoshimoto, J. V. Neel, W. J. Schull, H. Kato, M. Soda and R. Eto, 'Malignant tumors during the first two decades of life in the offspring of atomic bomb survivors' (1990) 46, *American Journal of Human Genetics*, 1041–52.

33 *op. cit.* note 2 at 27A. This and subsequent page references will refer to the transcript of the judgement of French, J. in the Queen's Bench Division of the High Court (8 October 1993).

34 M. Susser, 'The logic of Sir Karl Popper and the practice of epidemiology' (1986) 124, *American Journal of Epidemiology*, 711 at 713.

35 *ibid.* at 715.

36 *op. cit.* note 2 at 74B.

37 *ibid.* at 74D.

38 R. Doll, H. J. Evans and S. C. Darby, 'Paternal exposure not to blame' (1994) 367, *Nature*, 678–80.

39 I hesitate to speculate upon the capacity of a lay jury (in various US jurisdictions) to assimilate the information with a facility equal to that of Mr Justice French.

40 *Budden v. BP Oil* (1980) 124 Sol Jo. 376.

41 *op. cit.* note 13.

42 *op. cit.* note 9.

43 Ministry of Agriculture, Fisheries and Food, 'Post Chernobyl sheep restrictions', press release, December 1994.

44 *op. cit.* note 1, para. 12.

45 'Council Directive of 15 July 1980 amending the Directives laying down the basic safety standards for the health protection of the general public and workers against the dangers of ionising radiation', Directive 80/836/Euratom [1980] OJ L246/11, as amended by Directive 84/467/Euratom [1984] OJ L265/4.

46 Secretary of State for the Environment, *Radioactive Waste Management* (Cmnd 8607, 1982, HMSO).

47 Ionising Radiations Regulations 1985 (SI 1985, no. 1333).

48 International Commission on Radiological Protection, 'Recommendations of the ICRP' (1991) 21 (3) *Annals of ICRP*.

49 *op. cit.* note 1 at para 74.

50 *ibid.* at para 75.

51 National Radiological Protection Board, *Cost-benefit Analysis in Optimising the Radiological Protection of the Public: A Provisional Framework*, ASP4 (1981, HMSO).

52 National Radiological Protection Board, *Cost-benefit Analysis in the Optimisation of Radiological Protection*, ASP9 (1986, HMSO).

53 *Edwards v. National Coal Board* [1949] 1 KB 704.

54 Convened jointly under s.34 (1) of the Electricity Act 1957 and s.40 of the Town and Country Planning Act 1971.

55 Secretary of State for Energy, statement made on 11 June 1982 under rule 5 of *Electricity Generating Stations and Overhead Lines (Inquiries Procedures) Rules 1981* (SI 1981 no. 1841).

56 The cost of electromagnetic filters to remove radioactive crud from the coolant was calculated by the CEGB to exceed the 'cost' of the collective dose reduction which they might achieve.

57 Sir Frank Layfield, *Report of the Inquiry into Sizewell B PWR* (1987, HMSO).

58 *ibid.*, Chapter 35, paras 7, 35.

59 *op. cit.* note 5.

60 Health and Safety Executive, *HM Nuclear Installations Inspectorate: Safety Assessment Principles for Nuclear Power Reactors* (1979, HMSO) 96.

61 *Associated Provincial Picture Houses Ltd v. Wednesbury Corporation* [1948] 1 KB 223; [1947] 2 All ER 680.

62 *op. cit.* note 5 at 96.

63 M. Purdue, R. Kemp and T. O'Riordan, 'The Layfield Report on the Sizewell B Inquiry' [1987] *Public Law*, 162.

64 *op. cit.* note 57, Chapter 109, 14.

65 Health and Safety Executive, *The Tolerability of Risk from Nuclear Power Stations* (1988, HMSO) para. 3.

66 *ibid.*, Annex B, para. 26.

67 Health and Safety Executive, *Safety Assessment Principles for Nuclear Power Reactors* (1992, HMSO).

68 *ibid.*, para. 42.

69 *ibid.*, para. 80.

70 *ibid.*, para. 29.

71 *op. cit.* note 53.

72 *op. cit.* note 61.

73 *op. cit.* note 48, para. 112.

74 *ibid.*, para. 115.

75 *op. cit.* note 3; subsequent references indicate the page of the approved transcript of the judgement of Otton, J. in the Queen's Bench Division of the High Court, 29 September 1993.

76 Sir Roger Parker, *The Windscale Inquiry: Report by the Hon. Mr Justice Parker* (1978, HMSO).

77 *Town and Country Planning (Windscale and Calder Works) Special Development Order 1978* (SI 1978, no. 523).

78 *op. cit.* note 75, 35C.

79 *R. v. The Secretary of State for the Environment, HM Inspectorate of Pollution and the Minister of Agriculture, Fisheries and Food (ex parte Greenpeace and Lancashire County Council)* (1994) 6 *JEL* 297, [1994] 4 All ER 352; subsequent references indicate the page of the approved

transcript of the judgement of Potts, J. in the Queens Bench Division of the High Court, 4 March 1994.

80 *op. cit.* note 46.

81 Secretary of State for the Environment *et al.*, *Radioactive Waste: The Government's Response to the Environment Committee's Report*, (Cmnd 9852 1986, HMSO).

82 The Direction given in August 1977 to NRPB by the Secretary of State for Social Services refers to a duty to advise whenever there arise obligations, relating to radiological protection, to the European Communities, the Organisation for Economic Cooperation and Development, or an Agency of the United Nations.

83 Department of the Environment, *Radioactive Substances Act 1960: A Guide to the Administration of the Act* (1982, HMSO).

84 *op. cit.* note 79, 23D.

85 *op. cit.* note 45.

86 *op. cit.* note 79, 30E.

87 *op. cit.* note 79, 18G.

88 Department of the Environment, *Decision by the Secretary of State for the Environment and the Minister of Agriculture, Fisheries and Food in Respect of an Application from British Nuclear Fuels for Authorisations to Discharge Radioactive Wastes from the Sellafield Site* (1993, DOE).

89 *op. cit.* note 79, 45C.

90 T. O'Riordan, 'Prospects for the nuclear debate in the UK', in A. Blowers and D. Pepper (eds) *Nuclear Power in Crisis* (1987, Croom Helm) 295 at 299.

91 D. Franks, *Report of the Committee on Administrative Tribunals and Inquiries*, (Cmnd 213, 1957, HMSO).

92 For a discussion, see Chapter 3 of T. O'Riordan, R. Kemp and M. Purdue, *Sizewell B: An Anatomy of the Inquiry* (1988, Macmillan).

93 'Radwaste policy in tatters as Gummer blocks Nirex dump' (1997) 266, *ENDS Report*, 13.

94 *op. cit.* notes 46 and 81; and most recently, Secretary of State for the Environment, *Review of Radioactive Waste Management Policy – Final Conclusions*, Cmnd 2919 (1995, HMSO).

95 *op. cit.* note 93.

96 Department of Trade and Industry and Scottish Office, *The Prospects for Nuclear Power in the UK* (Cmnd 2860, 1995, HMSO).

CHAPTER 7

1 United Nations General Assembly, *Universal Declaration of Human Rights* (1948, New York, UNO).

2 By contending that it is only through the owner's labour in cultivating his land that it becomes fruitful, John Locke, *Two Treatises on Government* [1689] (1988, Cambridge University Press).

3 Communication from the Commission to the Council and Parliament and the Economic and Social Committee, *Green Paper on Remedying Environmental Damage*, COM (93) 47, 14 May 1993.

4 *ibid.*, para. 1.0.

5 J. Steele, 'Remedies and remediation: foundational issues in environmental liability' (1995) 58, *MLR*, 615.

6 UK Government, *Response to the Communication from the Commission of the European Communities COM (93) 47 final: Green Paper on Remedying Environmental Damage*, 8 October 1993, para 2.2.

7 *ibid.*

8 UK Government, *This Common Inheritance: Britain's Environmental Strategy* (Cm 1200, 1990, HMSO) para. 1.24, cited in note 6 at para. 3.2.

9 *op. cit.* note 6 at para. 3.2.

10 Council of Europe, *Convention on Civil Liability for Damage Resulting from Activities Dangerous to the Environment* (adopted 9 March 1993), Article 18.3.

11 *op. cit.* note 6 at para. 3.9.

It is necessary to locate this unequivocal assertion within the chronology of important rulings on the standing of public interest groups. It came three years after *Rose Theatre Trust* [1990] 1 QB 504 and one year after *Twyford* [1992] *JEL* 274 (discussed in Chapter 3 above), where the claims of the respective applicants to standing were rejected; but the most important precedent of a UK environmental group being permitted to present its (ultimately unsuccessful, see Chapter 6) case occurred only nine days before this response was published. Mr Justice Otton held in *Greenpeace (no. 2)* [1994] 4 All ER 329, that this organisation was 'eminently respectable and responsible and their genuine interest in the issues raised is sufficient for them to be granted *locus standi*' in the High Court on 29 September 1993. The importance of this ruling has been repeatedly stressed in the ensuing years; but it is hard to believe that its impact could have been quite so immediate as to have influenced the author of HM Government's memorandum (*op. cit.* note 6 above) which is dated 8 October 1993.

12 *op. cit.* note 6 at para 3.26.

13 For example, the International Convention on the Establishment of an International Fund for Compensation for Oil Pollution Damage (adopted in 1971) provides for compensation in excess of the limit, of liability for any one marine pollution incident, set by the International Convention on Civil Liability for Oil Pollution Damage 1969 (as amended by a Protocol of 1976). For a discussion of the limited liability of the operators of nuclear installations, see Chapter 6.

14 House of Lords Select Committee (1993–9) Session 3rd Report, *Remedying Environmental Damage* (14 December 1993).

15 Department of the Environment, *Government Response to 'Remedying Environmental Damage': Report by the House of Lords Select Committee* (4 May 1995).

16 Royal Commission on Environmental Pollution, *Eleventh Report: Managing Waste: The Duty of Care* (Cmnd 9675, HMSO, 1985) para. 9.31.

17 Located near Niagara Falls, New York State, Love Canal was an abandoned canal in which some 21,000 tons of chemical waste had been tipped between 1942 and 1952. In the late 1970s, Love Canal became increasingly notorious as miscarriages, birth defects, nervous breakdowns and other forms of morbidity were attributed to the effects of the leaking wastes. But a careful examination of the chronology will reveal that the legislative process that ultimately led to CERCLA and Superfund was already under way when concern over this particular tip was still confined to the local community. See M. K. Landy, M. J. Roberts and S. R. Thomas, *The Environmental Protection Agency: Asking the Wrong Questions* (1990, New York, Oxford University Press).

18 *Civil Justice Reform* [Executive Order 12, 988, 61 Fed. Reg. 4729 (7 February 1996)] made alternative dispute resolution a 'priority' across the federal government.

19 Lord Woolf, *Access to Justice: Final Report to the Lord Chancellor on the civil justice system in England and Wales* (1996, HMSO).

20 *United States v. Fleet Factors Corporation*, 901 F.2d 1550 (11th Cir. 1990).

21 *op. cit.* note 14, para. 83.

22 The UK government response (*op. cit.* note 6 above) to the EC green paper on remedying environmental damage makes only passing reference to the 'unowned environment' and argues that UK law empowers regulatory authorities adequately to protect it; but it declines to cite any examples of recipients of this protection.

23 J. E. Bonine and T. O. McGarity, *The Law of Environmental Protection*, 2nd edn (1992, Minnesota, West Publishing) 907.

24 Directive 80/778/EEC [1980] OJ L229/11.

25 *Cambridge Water Company v. Eastern Counties Leather plc* [1994] 2 WLR 53.

26 (1866) L.R. 1 Ex. 265, affirmed (1868) L.R. 3 H.L. 330. at 280.

27 Directive 80/68/EEC [1980] OJ L20/43.

28 D. Wilkinson, '*Cambridge Water Company v. Eastern Counties Leather plc*: diluting liability for continuing escapes' (1994) 57, *MLR*, 799.

29 *op. cit.* note 10.

30 s.78A(2), Environmental Protection Act 1990 Act as amended.

31 Department of the Environment, *Draft Guidance on Determination of Whether Land is Contaminated Land Under the Provisions of Part IIA of the Environmental Protection Act 1990*, 5 May 1995.

32 *op. cit.* note 30, s.78B.

33 In the case of sites posing special technical problems, the Environment Agency, under regulations yet to be issued, assumes the regulatory role in place of the district council.

34 The term 'occupier' is not defined; 'owner' is defined (s.78A of the 1990 Act) so as to exclude mortgagees not in possession and insolvency practitioners unless the contamination resulted from their negligence.

35 *op. cit.* note 31, para 3.

36 Directive 75/442/EEC [1975] OJ L194/39 as amended by Directive 91/156/EEC [1991] OJ L78/32.

37 See N. Haigh, *EEC Environmental Policy and Britain* (2nd edn, 1987, Longman) 134.

38 *op. cit.* note 36, Article 4.

39 Case C-236/92, *Comitato de Coordinamento per la Difesa della Cava and others v. Regione Lombardia and others* [1994] ECR I-485.

40 See analysis by J. Holder, 'A dead end for direct effect?: prospects for enforcement of European Community Environmental law by individuals' (1996) 8 *JEL* 313 at 330.

41 County councils in England, district councils in Wales. Part I of the Control of Pollution Act 1974 has been effectively replaced by Part II of the Environmental Protection Act 1990, and enforcement has passed to the Environment Agency, leaving the district councils as the 'waste collection authorities' overseeing the collection of waste by private contractors.

42 Control of Pollution (Special Waste) Regulations 1980 (SI 1980, no. 1709).

43 Special Waste (Amendment) Regulations 1996 (SI 1996 no. 2019).

44 (1993) 216, *ENDS Report*, 43.

45 Methane gas from a landfill was responsible for an explosion which demolished a bungalow in the Derbyshire village of Loscoe in 1986. Personal injuries suffered in a similar explosion in a Toronto garage had earlier led to a successful action in *Rylands v. Fletcher*, nuisance and negligence, *Gertsen v. Municipality of Toronto* [1973] 41 DLR (3d) 646. There does not appear to have been a reported case of comparable importance within UK jurisdiction despite a plethora of incidents. In an out-of-court settlement, the insurers of a landfill business are reported to have paid £750,000 to a building firm unable to complete a housing development on account of the methane (observed bubbling through puddles) alleged to originate from the landfill. This claim was originally disputed by the plaintiffs when their carbon dating of the gas pointed to local coal seams as the source (1989) 168, *ENDS Report*, 7.

46 Secretary of State for the Environment, *Making Waste Pay: A Strategy for Sustainable Waste Management in England and Wales*, (Cm. 3040, 1995, HMSO).

47 Under the 1989 Electricity Act a number of renewable energy and waste-to-electricity projects were funded from a levy on all electricity users.

48 Conceivably by restrictive conditions, contained in regulations made by the Secretary of State under s.35(6) of the 1990 Act, appended to the waste management licences of all landfills and similarly unfavoured disposal options. Now that the Environment Agency has replaced the various waste regulation authorities, the pursuit of policy objectives by administrative measures rather than by statutory instrument becomes far easier.

49 Directive 94/67/EEC [1994] OJ L365/34.

50 HM Customs and Excise, *Landfill Tax – A Consultation Paper* (March 1995).

51 Royal Commission on Environmental Pollution, *Nineteenth Report: Sustainable Use of Soil* (Cm. 3165, HMSO, 1996) para. 10.61.

52 Trevor Lawson, 'Danger of being dumped', *Guardian*, 1 October 1997.

53 Robert Lewis, 'Contaminated land: the new regime of the Environment Act 1995' [1995] JPL 1087.

54 In fact, the Royal Commission advocated the extension of this duty from waste disposal to environmental protection in general (*op. cit.* note 16 at para 3.5). It was seen as analogous to employers' duty (under s.2 of the Health and Safety at Work Act 1974) in regard to the health, safety and welfare of their employees.

55 Harry Barton, 'The Isle of Harris superquarry: concepts of the environment and sustainability' (1995) 5, *Environmental Values*, 97.

CHAPTER 8

1 Rachel Carson, *Silent Spring* (1982, Penguin)

2 Marion Shoard, *The Theft of the Countryside* (1980, Temple Smith)

3 Aldo Leopold, *A Sand County Almanack* (1949, Oxford University Press)

4 This '1949 Act', together with the Town and Country Planning Act 1947 and the New Towns Act 1946, formed the principal elements of the framework of town and country planning introduced by the postwar Labour Government and still largely intact today.

5 Access agreements, between local planning authorities and the owners of specified areas of land within national parks, confer an immunity from trespass, s.60 of the 1949 Act.

6 s.114.2, 1949 Act.

7 ss.15–22, 1949 Act.

8 I use the generic term 'conservancy council' to refer to the Nature Conservancy Council, established in its final form by the Nature Conservancy Act 1973 and responsible for nature conservation throughout Great Britain, and (since 1991, following changes contained in the Environmental Protection Act 1990) to its three successors, namely the Nature Conservancy Council for England, the Countryside Council for Wales and Scottish Natural Heritage. Each of these three councils consists of about twelve members all appointed by the relevant Secretary of State. Coordination and a common voice on issues of shared concern are attempted by a Joint Nature Conservation Committee. Since 1991 the Countryside Commission's remit is confined to England.

9 Directive 79/409/EEC [1979] OJ L103/1.

10 Directive 92/43/EEC [1992] OJ L206/7.

11 s.28, Wildlife and Countryside Act 1981.

12 *ibid.*, s.42.

13 *ibid.*, s.39.

14 *ibid.*, s.28.

15 ss.16–17, 1949 Act.

16 Department of the Environment, *Wildlife and Countryside Act 1981: Financial Guidelines for Management Agreements*, Circular 4/83 (1983, HMSO).

17 s.50, 1981 Act.

18 [1992] 3 All ER 481.

19 s.28, 1981 Act.

20 *op. cit.* note 18 at 485.

21 P. Lowe, G. Cox, M. MacEwen, T. O'Riordan and M. Winter, *Countryside Conflicts: The Politics of Farming, Forestry, and Conservation* (1986, Gower).

22 Now contained in s.31(2), Town and Country Planning Act 1990. This could be interpreted as a particular instance of the general duty imposed, by s.11 of the Countryside Act 1968, on all ministers, government departments and public bodies to take account of the desirability of conserving the beauty and amenity of the countryside when exercising their various functions.

23 Conservation (Natural Habitats etc.) Regulations 1994 (SI 1994, no. 2716).

24 *op. cit.* note 10.

25 Department of the Environment, *Development Control – Policy & Practice*, Circular 22/80 (1980, HMSO) para. 4.

26 Nigel Curry, 'Controlling development in the national parks of England and Wales' (1992) 63, *Town Planning Review*, 107–21.

27 The Town and Country Planning (Assessment of Environmental Effects) Regulations 1988 (SI 1988, no. 1199), schedule 2 lists those categories of development for which environment assessment is not mandatory but dependent upon scale.

28 Department of the Environment, *Environmental Assessment*, Circular 15/88 (1988, HMSO).

29 *R. v. Poole Borough Council ex parte Beebee* [1991] JPL 643.

30 *op. cit.* note 27.

31 John Alder, 'Environmental impact assessment – the inadequacies of English law' (1993) 5 *JEL* 203 at 213.

32 *ibid.*

33 *R. v. Swale Borough Council and the Medway Ports ex parte the Royal Society for the Protection of Birds* (1991) 3 *JEL* 135.

34 A. Ward, 'The right to an effective remedy in European Community law: a case study of United Kingdom judicial decisions concerning the Environmental Assessment Directive' (1993) 5 *JEL* 221 at 229.

35 *op. cit.* note 27.

36 *op. cit.* note 33 at 143.

37 *op. cit.* note 10.

38 There were 280 land-based SACs notified to Commission in June 1995, [1995] *JPL*, 499. A further 75 (10 of which were marine) were notified in February 1996, [1996] *JPL*, 199.

39 *op. cit.* note 23.

40 Town and Country Planning (General Permitted Development) Order 1995 (SI 1995 no. 418) lists in Schedule 2 those of 'agricultural buildings and operations' which normally constitute 'permitted development'. But Article 3 makes it clear that this designation is contingent upon the Regulations 60–3 of the Habitats Regulations 1994.

41 Case C-57/89, *Commission v. Federal Republic of Germany* [1991] ECR I-883.

42 *ibid.*

43 Case C-355/90, *Commission v. Spain* [1993] ECR I-4221.

44 *R v. Secretary of State for the Environment ex parte the Royal Society for the Protection of Birds* (1995) 7 *JEL* 245.

45 Case C-44/95, *R. v. Secretary of State for the Environment ex parte the Royal Society for the Protection of Birds* (1997) 9 *JEL* 139.

46 Advocate General's opinion given 21 March 1996, *ibid.*

47 *ibid.* at 166.

48 *ibid.* at 167.

49 *op. cit.* note 21 at 296.

50 *op. cit.* note 40.

51 I am indebted to Professor C. P. Rogers for this point.

52 *op. cit.* note 41.

53 Department of the Environment, *Planning Policy Guidance: Nature Conservation*, PPG 9 (1994, HMSO).

54 (1991) 194, *ENDS Report*, 32.

55 Royal Society for the Protection of Birds, *Possible Special Areas of Conservation (SACs) in the UK* (1995, Sandy, RSPB).

56 English Nature, *Fourth Report* (1995, Peterborough, English Nature).

57 Council Regulation 797/85 [1985] OJ L93/1 on improving the efficiency of agricultural structures.

58 N. Haigh, *EEC Environmental Policy and Britain* (2nd edn, 1987, Longmans) 311.

59 Council Regulation 2328/91 [1991] OJ L218/1.

60 House of Lords Select Committee on the European Communities, *Environmental aspects of the reform of the common agricultural policy* (Session 1992–3, 14th Report, HMSO) 15, written evidence from CPRE, para. 15.

61 The Nitrate Sensitive Areas (Designation) Order 1990 (SI 1990, no. 1013) under powers then contained in s.112 of the Water Act 1989.

62 In regulations made under s.2(2) of the European Communities Act 1972 and not, it should be noted, under the pollution control provisions of a statute concerned with water (see notes 61 above and 64 below).

63 Directive 91/676/EEC [1991] OJ L375/1.

64 Under ss.94–5, Water Resources Act 1991.

65 *op. cit.* note 60.

66 *op. cit.* note 46.

67 *op. cit.* note 41.

68 D. W. Pearce, A. Markandya and E. B. Barbier, *Blueprint for a Green Economy* (1989, London, Earthscan).

69 M. Freeden, *Rights* (1991, Milton Keynes, Open University Press) 7.

70 R. Dworkin, *Taking Rights Seriously* (1977, London, Duckworth).

CHAPTER 9

1 *Sierra Club v. Morton* 401 US 907 (1971).

2 Christopher Stone, *Should Trees Have Standing? Towards Legal Rights for Natural Objects* (1974, Los Altos CA).

3 Case C-44/95 *R. v. Secretary of State for the Environment ex parte the Royal Society for the Protection of Birds* (1997) 9 *JEL* 139.

4 Although the European Court of Justice ruled in RSPB's favour, the victory was somewhat pyrrhic since the House of Lords had earlier declined to grant the injunction that would have halted engineering works on Lappel Bank which, in the event, were completed whilst Luxembourg deliberated.

5 A similar historical irony arises when conservationists rely upon various eighteenth-century local acts of Parliament to prevent the destruction of hedgerows; see Martin Wainwright 'Saved (for now): the great British hedge', *Guardian*, 3 January 1997. These hedgerows were once the most visible and hated signs of the enclosure of the rural commons and the extinction of the rights previously enjoyed by occupants (tenants) with no other claim upon the land. According to one of the leading social historians of this period: 'Enclosure (when all the sophistications are allowed for) was a plain enough case of class robbery, played according to the fair rules of property and law laid down by a parliament of property-owners and lawyers' (E. P. Thompson, *The Making of the English Working Class*, 1963, Penguin) 237.

6 *League against Cruel Sports Ltd v. Scott and others* [1985] 2 All ER 489.

7 s.120(1), Local Government Act 1972.

8 'Hunters scent blood over stag ban', *Independent*, 22 August 1997.

9 *R. v. Somerset County Council ex parte Fewings* [1995] 1 All ER 513.

10 Yongo, T, '*Fewings* in the Court of Appeal' (1994–5) 1, *EJRB*, 42.

11 'Wergeld', under Anglo-Saxon law, seems the closest (albeit anthropocentric) analogue.

12 The Supreme Court held that the slaughter of a bear (protected by national legislation implementing the Berne Convention on the Conservation of European Wildlife and Natural Habitats) violated a legal good belonging not only to the Asturian Fund but to society as a whole and was not to be considered 'as a mere patrimonial loss' [1993] *Environmental Liability*, CS77.

13 s.3 of the Children Act 1989 defines 'parental responsibility' (whether vested in a natural parent, guardian or a local authority) as 'all the rights, duties, powers, responsibilities and authority which by law a parent of a child has in relation to the child and his property'.

14 s.1(1), Children Act 1989.

15 See Emile Durkheim, *The Rules of Sociological Method* (1938, New York, Free Press).

16 My choice of the word 'similar' rather than 'analogous' is deliberate.

17 Neil MacCormick, *Legal Right and Social Democracy* (1982, Oxford University Press) 155.

18 Simon Hattenstone, 'Little earners', *Guardian*, 3 March 1993.

19 Simon Lyster, *International Wildlife Law* (1993, Cambridge University Press) 23.

20 Schedule 8, Wildlife and Countryside Act 1981.

21 Thomas Nagel, 'What is it like to be a bat?' in *Mortal Questions* (1979, Cambridge MA, Harvard University Press) 168.

22 *op. cit.* note 1.

23 Richard Dawkins, *The Selfish Gene* (1976, Oxford University Press).

24 It is worth noting that the notion of compensation does appear in Article 6.4 of the EC Habitats Directive (92/43/EEC [1992] OJ L206/7) where 'compensatory measures necessary to ensure the overall coherence of Natura 2000 is protected' must accompany any encroachment 'for imperative reasons of overriding interest' on a designated site.

25 *op. cit.* note 17 at 154.

26 Paola Cavalieri and Peter Singer, *The Great Ape Project: Equality beyond humanity* (1993, Fourth Estate).

27 Royal Commission on Environmental Pollution, *Nineteenth Report: Sustainable Use of Soil* (Cm. 3165, HMSO, 1996) para. 3.70.

28 Peter Singer, *Animal Liberation* (1975, New York, The New Review).

29 Tom Regan, *The Case for Animal Rights* (1983, Berkeley CA, University of California Press).

30 The Convention on International Trade in Endangered Species of Wild Fauna and Flora, see Simon Lyster, *International Wildlife Law* (1993, Cambridge University Press) 244.

31 Royal Commission on Environmental Pollution, *Thirteenth Report: The Release of Genetically Engineered Organisms to the Environment* (Cm. 720, 1989, HMSO).

32 John Rodman, 'The Liberation of Nature?' (1977) 20, *Inquiry*, 83–145.

33 Ted Benton, *Natural Relations: Ecology, Animal Rights and Social Justice* (1993, London, Verso).

34 Holmes Rolston III, 'Can and ought we to follow nature?' (1979) 1, *Environmental Ethics*, 7–30.

35 Warwick Fox, *Towards a Transpersonal Ecology: Developing New Foundations for Environmentalism* (1990, Boston MA, Shambala).

36 But with ecocentric (as with any other class of) rights, any correlative duties remain firmly grounded in the anthropocentric camp, because only adult, mentally capable humans can be the bearers of any form of duty, obligation or responsibility; see Onora O'Neill, 'Environmental values, anthropocentrism and speciesism' (1997) 6, *Environmental Values*, 127 at 133.

37 C. Glacken, *Traces on the Rhodian Shore* (1967, Berkeley CA, University of California Press).

38 L. White, 'The historical roots of our ecologic crisis' (1967) 155, *Science*, 1203.

39 See for example, J. Carp, 'A private property duty of stewardship: changing our land ethic' (1993) 23, *Environmental Law*, 735.

40 Joseph Raz, *The Morality of Freedom* (1986, Oxford University Press) 186.

41 *ibid.* at 212.

42 For instance, the 'direct effect' of European Community directives discussed at length in Chapter 2.

NAME INDEX

212

SUBJECT INDEX

Printed and bound in Great Britain by
TJ International Ltd, Padstow, Cornwall